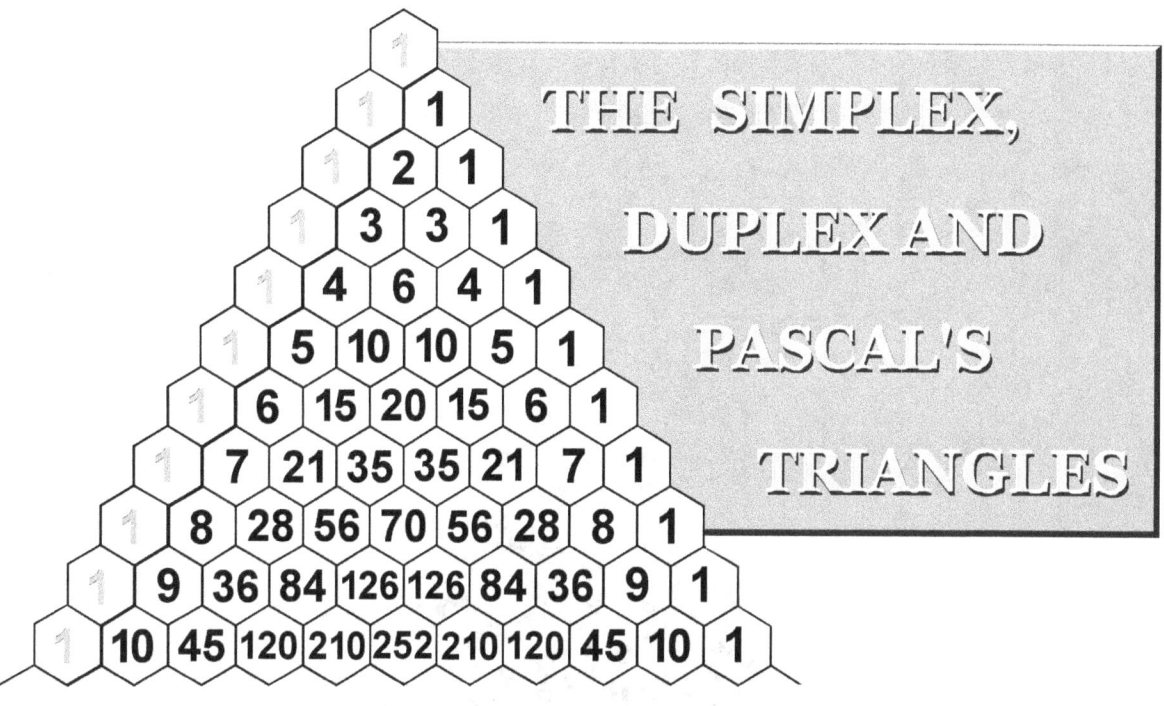

# THE SIMPLEX, DUPLEX AND PASCAL'S TRIANGLES

# WITH EXCURSIONS INTO HYPERSPACE

**Thomas M. Green**

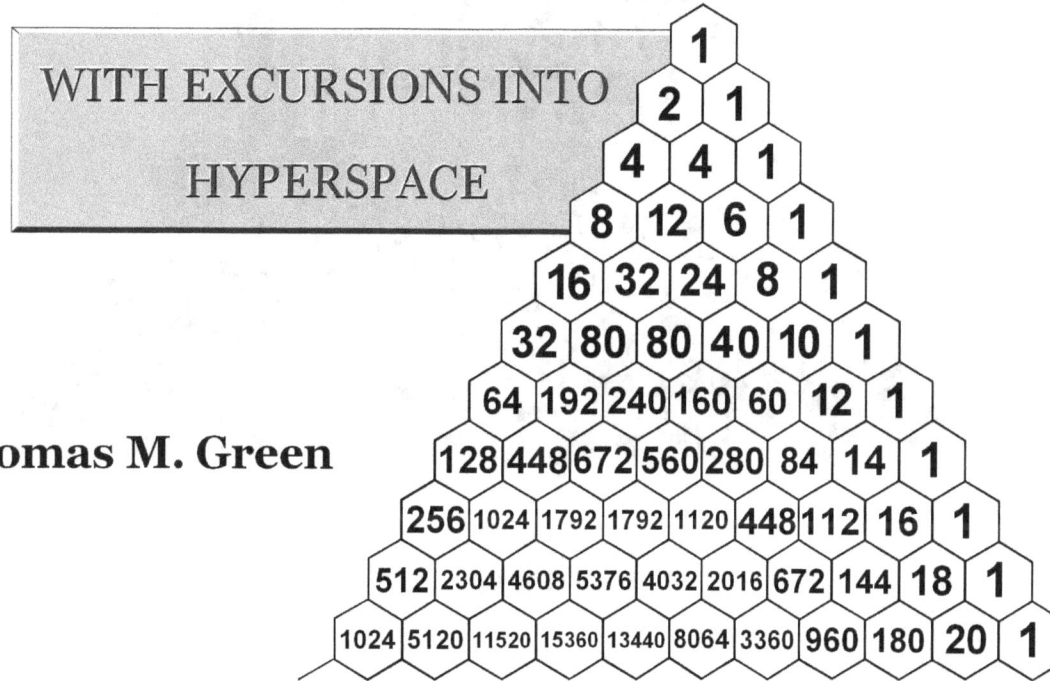

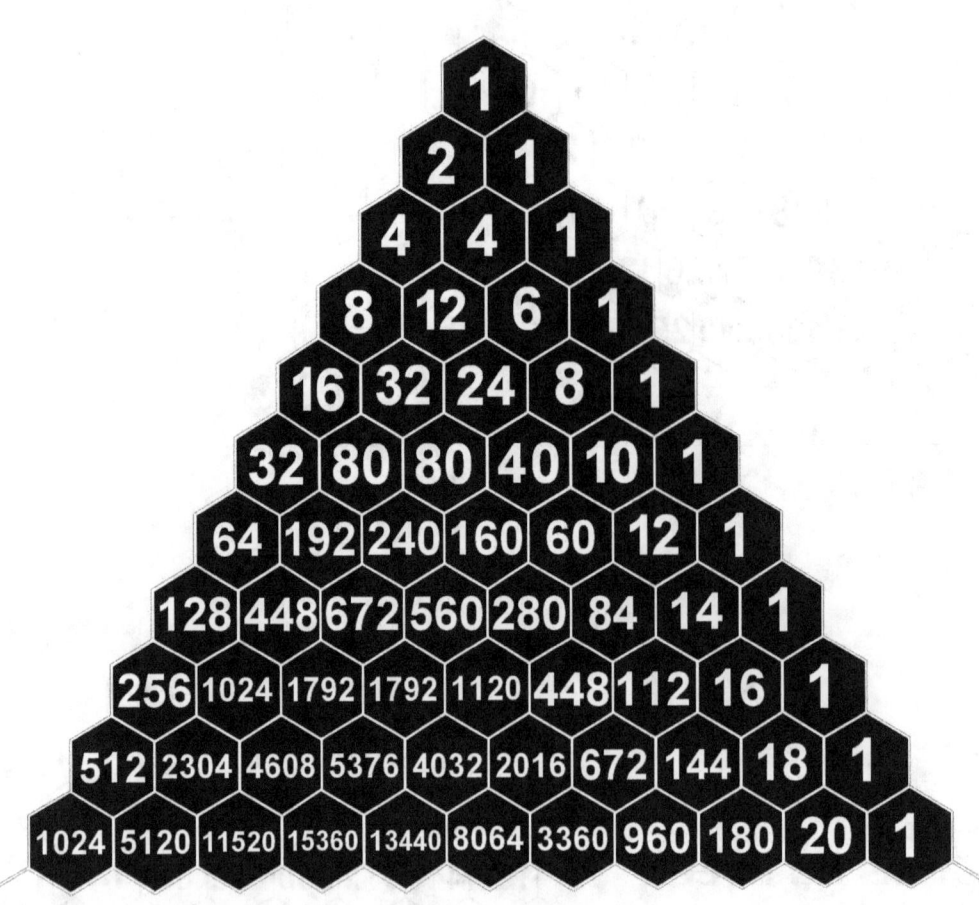

**DUPLEX TRIANGLE**

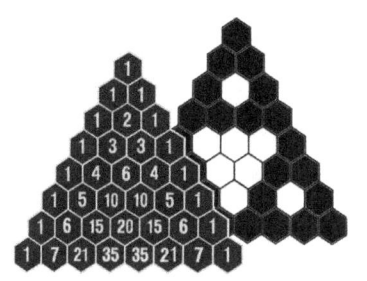

# The Simplex, Duplex,

# and

# Pascal's Triangles

### Relatives of Pascal's Triangle,

### A Family of Related Number Triangles

### With Excursions into Hyperspace

## THOMAS M. GREEN

RIGHT  ANGLE

## You are Invited to Visit

https://www.createspace.com/5578019

**For pricing and ordering information.**

This book is dedicated to my Wife, Sandra,
To all my other loved ones,
To friends,
To my ancestors,
And to my descendents, whoever they shall be.

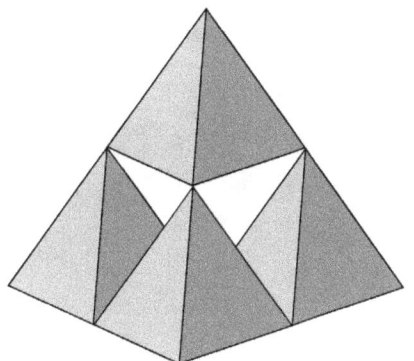

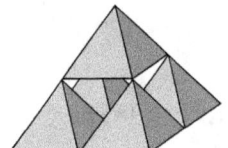

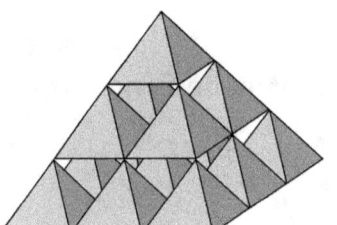

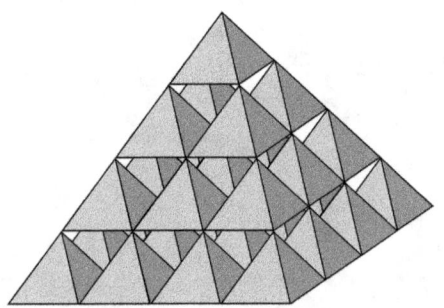

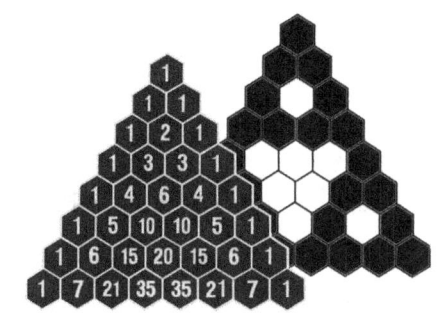

# Contents

## Chapter 6    Another Relative of Pascal's Triangle          **117**

## Chapter 7    A Family of Pascal Related Triangles          **153**

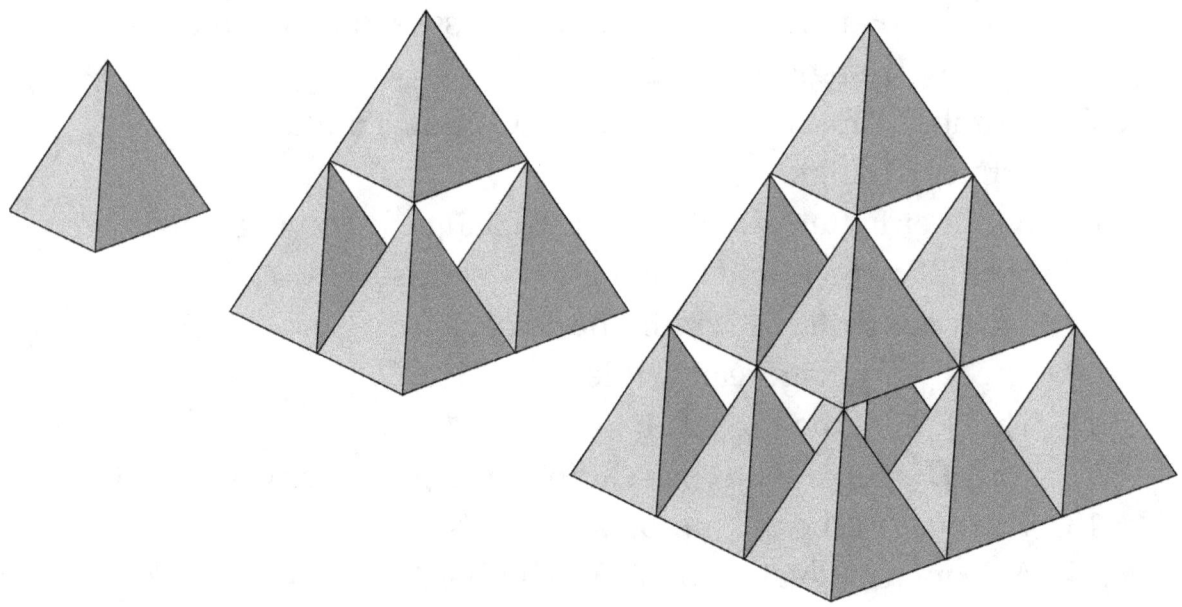

**T**etrahedrons stacked in the shape of tetrahedrons,
forming representations of the first three tetrahedral
numbers by count of the smaller parts, 1, 4, and 10.

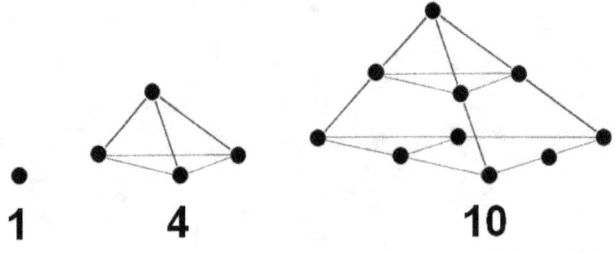

**1**      **4**       **10**

# Preface

Prepare to be intrigued by the many facets of the properties of the amazing array of numbers known as Pascal's Triangle and its many relatives. Some of the topics you will find:

Polytopes, Simplexes and the Simplex Triangle, triangular, tetrahedral, and higher dimensional figurate numbers, Duplexes, cubes and hypercubes, geometric duplication, the Duplex Triangle, Vandermonde's Identity for the Duplex Triangle and the Triplex Triangle, Euler's formula for simplexes and duplexes, recurrent sequences in Pascal's Triangle and its relatives, properties involving string products, Pythagorean triples, to name a few.

There is a comprehensive index that will allow readers to easily search for topics of their interest.

One goal is to provide a vehicle to the discovery of some higher mathematics related to higher dimensional geometric figures, at an entry level for the young beginning researcher by including many exercises that ask for verification of a pattern by testing specific cases and conjecturing a generalization of the pattern.

Another major goal was to make available source materials for mathematics teachers to use in their classes. Included are many topics suitable for introducing students, at the pre-college level, to the sense of satisfaction one receives while exploring and discovering significant parts of advanced mathematics.

I hope you will enjoy exploring this amazing Arithmetic Triangle and its relatives as much as I have. There is still much more to be discovered, of that I am certain.

The Author, 2015

*". . . I leave out many more [uses of the Arithmetic Triangle] than I include; it is extraordinary how fertile in properties this triangle is. Everyone can try his hand."*

*Blaise Pascal*
*"Treatise on the Arithmetic Triangle," 1653*

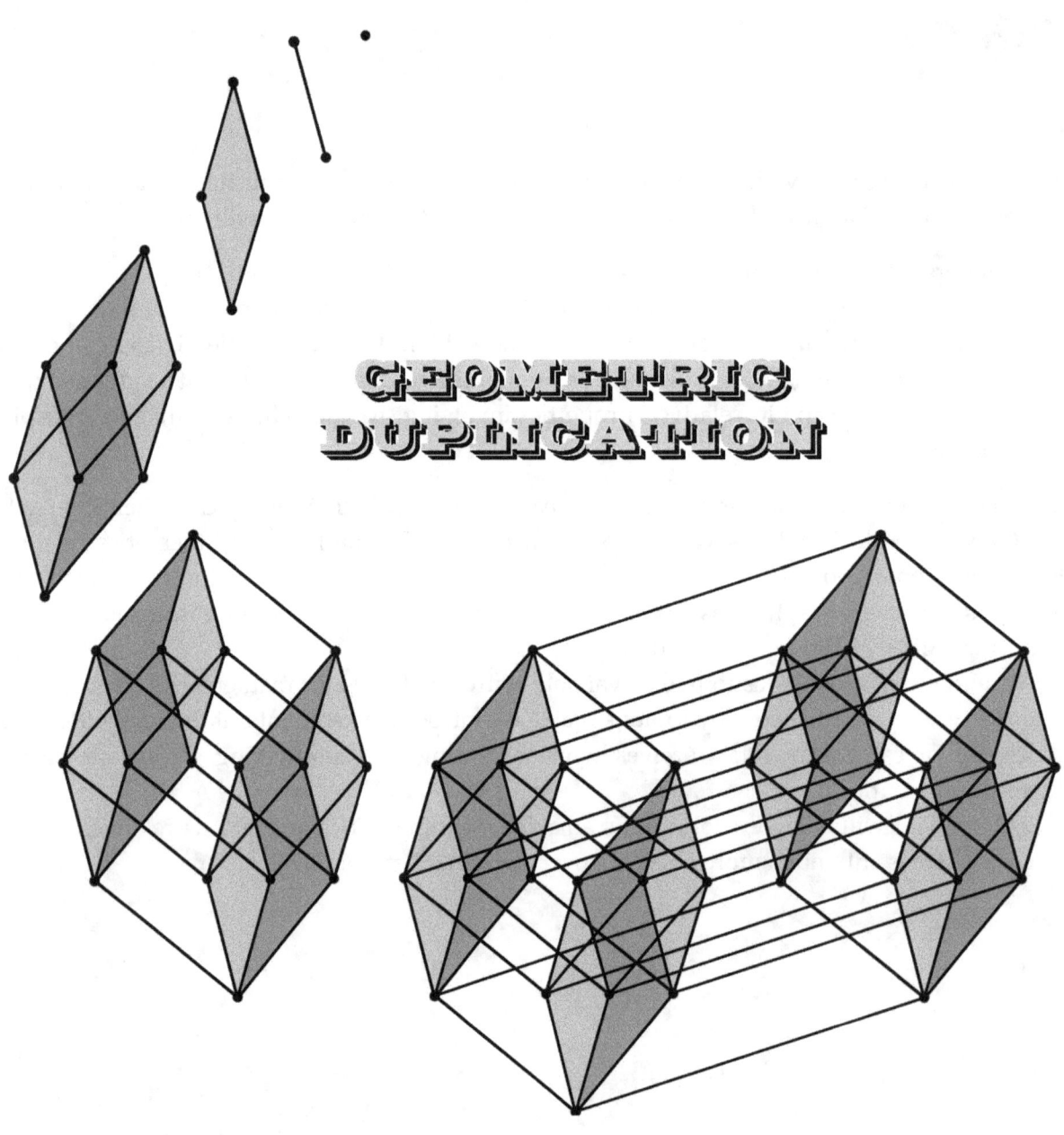

**The first 5 stages of geometric duplication showing duplexes (distorted) from 0-D to 5-D.**

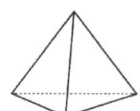 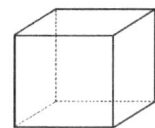 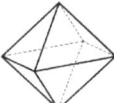 

# Introduction

## Polytopes

In the chapters to follow we will explore three strands of geometric figures that extend to an unlimited number of dimensions. Geometric figures of higher dimensions are based on mathematical concepts and not so much on physical reality. The general term for these figures is "polytope". The vocabulary and symbolism to describe and analyze geometrical space is varied and formidable, see [1] and [2]. H. S. M. Coxeter in his book "Regular Polytopes", presents a foremost exposition of the subject. He writes:

"**Polytope** is the general term of the sequence
point, segment, polygon, polyhedron, . . . ."

The dimensions of the first four elements in this sequence are 0, 1, 2, and 3, but the necessity of creating the term *polytope* is to refer to additional elements of the sequence when extended to geometric figures of higher dimensions.

Polytopes of a given dimension need not be *convex* nor *regular*, but they do consist of points, edges, faces, etc., that are all polytopes of lesser dimensions. Polytopes of dimension **d** have an interior and an exterior, and the interior is separated from the exterior by a **boundary** consisting of a collection of polytopes of dimension **d − 1**, each of which are bounded by polytopes of dimension **d − 2**, each of which are bounded by polytopes of dimension **d − 3**, and so on, until we reach dimension **0**, which are points called the vertices of the polytope. The simplest polytope that can be drawn in any dimensional space is called a **simplex**.

There are three specific strands of *regular convex polytopes* that extend to unlimited higher dimensions. The first of these strands is called the **simplex strand** and is presented in Chapter 1. The sequence of simplexes in this strand, starting with dimension **0** and in increasing dimensions, is point, followed by line-segment, triangle, tetrahedron, pentatope (the simplest figure in **4**-dimensions), and so on, forming a strand or sequence of geometric figures through **d**-dimensions. When we run out of vocabulary in naming a simplex of a higher dimension, **d**, we will refer to it as a **d**-simplex. The figures in this strand are *regular and convex*; the meaning of these terms is discussed in Chapter 1, but for now we will simply say that a regular and convex **2**-simplex is the familiar equilateral, or equiangular, triangle.

The second and third strands of *regular convex polytopes* we will call the **duplex strand** and the **reciprocal of the duplex strand**. Elements of the duplex strand of regular convex polytopes are point, line-segment, square, cube, tesseract (a hypercube, a figure in **4**-dimensions), and so on, forming the strand or sequence of polytopes through **d**-dimensions called "**duplexes**". This strand and its reciprocal strand are also introduced in Chapter 1.

*(Note: the term "duplex" is not widely used to name this strand of polytopes. Indeed, the term duplex is deeply ingrained in the English language to mean other things, so for the purposes of this text, we ask that you, the reader, temporarily suspend your understanding of the term as it is commonly used and adopt the meaning being presented here. Coxeter [1] refers to elements in this strand as "measure polytopes" and D. M. Y. Sommerville [2] refers to them as "regular orthotopes". Some authors use the term "hypercube" to refer to any higher dimensional figure in this strand, not just the $4^{th}$ dimension.*

We will see, in Chapter 1, the term "duplex" accurately describes this strand, since each element in the strand is formed from the one immediately preceding it by a process called geometric duplication (see p. xii, p. 4, and Chap 1). The term also fits nicely with the term simplex, and like that term, "duplexes", in general, refers to the entire strand of these particular polytopes.

The three pivotal elements in 3-D for each of the three strands of regular and convex polytopes are tetrahedron, cube, and octahedron.

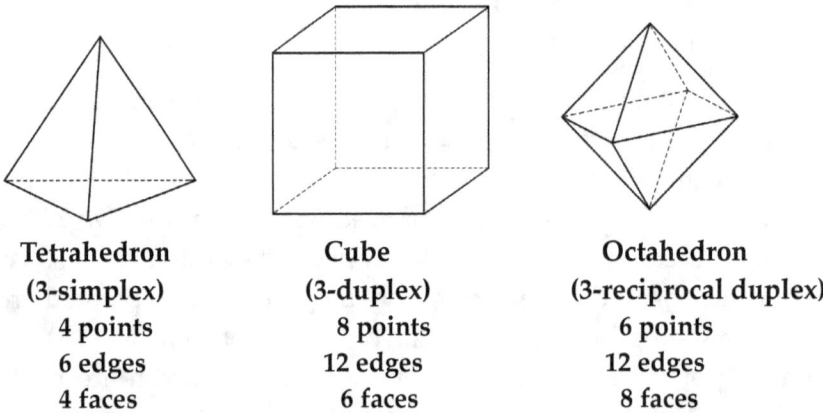

| Tetrahedron | Cube | Octahedron |
|:---:|:---:|:---:|
| (3-simplex) | (3-duplex) | (3-reciprocal duplex) |
| 4 points | 8 points | 6 points |
| 6 edges | 12 edges | 12 edges |
| 4 faces | 6 faces | 8 faces |

**Figure 1.   The 3 pivotal  3-D elements within each strand of regular convex polytopes, listing the type and quantity of sub-elements of the polytope.**

In Chapter 1 we also explore the lengths of the diagonals of the duplex figures, including the hypercube and higher dimensional duplexes. There is also an interesting generalization of the Pythagorean Theorem, relating the edges and the diagonals of duplexes, namely,

*The sum of the squares of the lengths of all the edges of a **d**-duplex*
*is equal to*
*the sum of the squares of  the lengths of all of its diagonals.*

## Number Triangles

Closely associated with the three strands of regular and convex polytopes are three different two dimensional arrays of numbers (shaped in a triangular arrangement), and each is related to Pascal's Triangle. These are called the **Simplex Triangle**, **Duplex Triangle**, and the **Reciprocal of the Duplex Triangle**.

A given row of numbers in these arrays corresponds to a specific dimension of a given polytope. The numbers in the row describe the type (by position) and quantity (by number) of sub-elements contained in the polytope. These number triangles and their additional properties and applications are developed in the remaining chapters of the book. They also give rise to other Lucas type sequences, namely, the **Fibonacci** sequence, the **Pell** sequence, the **Jacobsthal** sequence, and others. The numbers in these sequences share a **Pythagorean connection** which is covered in each Chapter.

**Rows 0 through 4 of the Simplex Triangle**

**Rows 0 through 4 of the Duplex Triangle**

**Rows 0 through 4 of the Reciprocal Duplex Triangle**

In Chapter 6 we develop an extension to the sequence of number triangles, called the **Triplex Triangle**.

## Note to the Reader

If you are familiar with elementary algebra, geometry and Pascal's Triangle, and you have an interest in mathematics and some imagination, you will be able to fully appreciate the concepts and ideas presented within. If you are not familiar with Pascal's Triangle, we recommend Pascal's Triangle, 2nd Ed., [3].

Hopefully, along the way you will find something new and exciting for you, and/or, perhaps you will discover something else related to the ideas and concepts presented here. If you do, please share it with others.

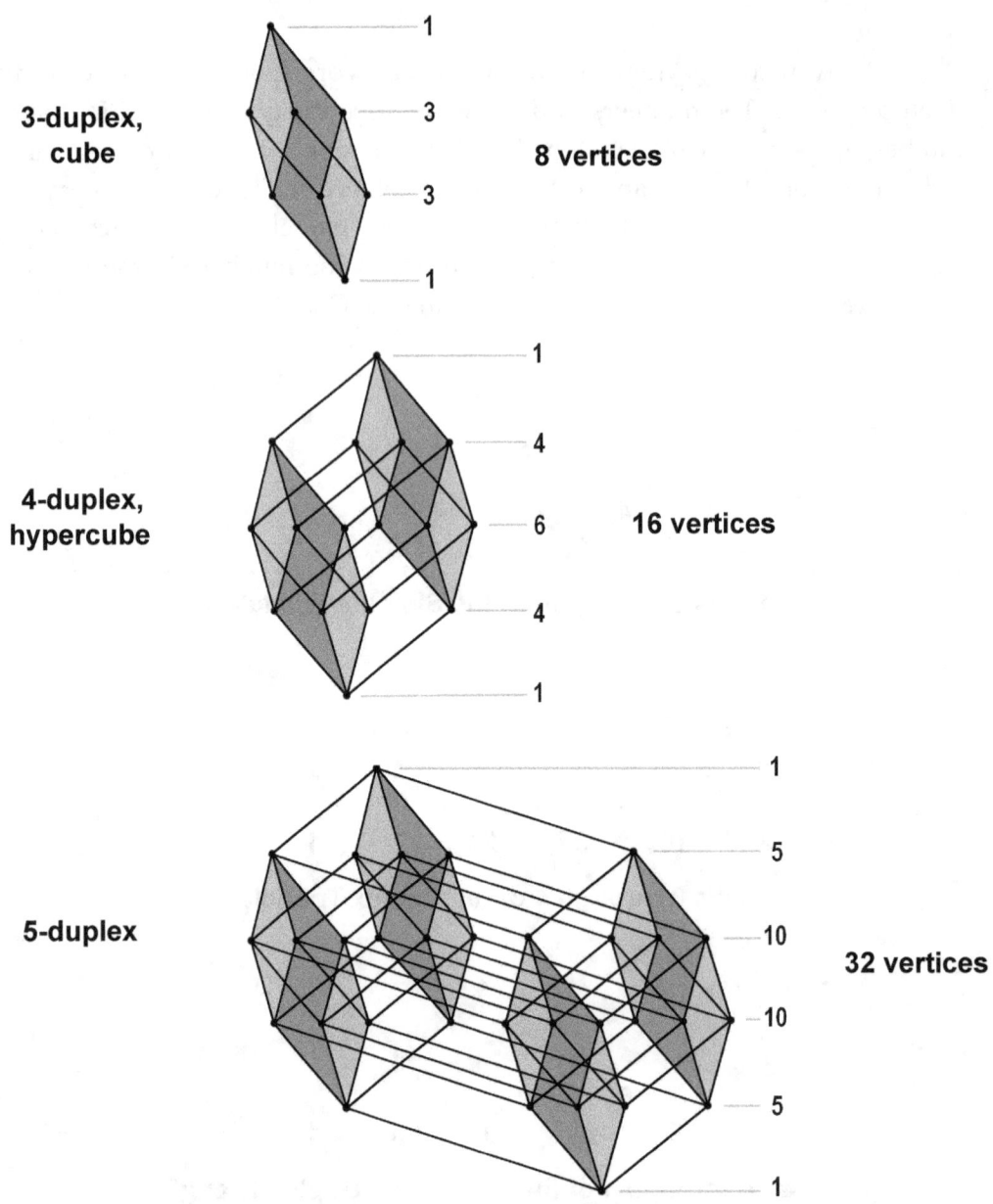

**3-duplex, cube**     1, 3, 3, 1     **8 vertices**

**4-duplex, hypercube**     1, 4, 6, 4, 1     **16 vertices**

**5-duplex**     1, 5, 10, 10, 5, 1     **32 vertices**

Figure 2.   Counting the vertices of figures in the Duplex strand of regular convex polytopes,
for  d = 3, 4, and 5,   see Chapter 1 (also see [3] ).
(The figures shown here have distorted angles and lengths of edges, but parallelism is preserved.)

### References

1.   H. S. M. Coxeter, *Regular Polytopes*, *3ʳᵈ Edition*, Dover Publications, Inc., 1973.

2.   D. M. Y. Sommerville, *Introduction to the Geometry of N Dimensions*, Methuen & Co. Ltd.,
UK, 1929.
(A PDF copy of this book can be downloaded for free from Archive.org.  Use the link
https://archive.org/details/IntroductionToTheGeometryOfNDimensions

3.   Thomas M. Green  and  Charles L. Hamberg, *Pascal's Triangle*, *2ⁿᵈ Edition*, Createspace,
2012.

# 1

# The Three Strands of Regular Convex Polytopes

## 1.1 Regular Convex Polytopes

The term "polytope", unknown in mathematical literature until the the late 19th Century, was first introduced by Reinhold Hoppe in Germany in 1882 and reintroduced to England by Alicia Boole Stott, daughter of the English mathematician George Boole, in the early 20th century. She became an aquaintenance of H. S. M. Coxeter in 1930, see [1] p. 259, who writes

"**Polytope** is the general term of the sequence

point, segment, polygon, polyhedron, . . . ." See [1], p. 118.

The dimensions of the elements in this sequence are 0, 1, 2, 3, . . . .

There are three strands of regular convex polytopes that extend to unlimited higher dimensions. We will call the three strands the **simplex strand**, the **duplex strand** and the **reciprocal of the duplex strand**. The three pivotal elements in 3-D for each strand are tetrahedron, cube, and octahedron.

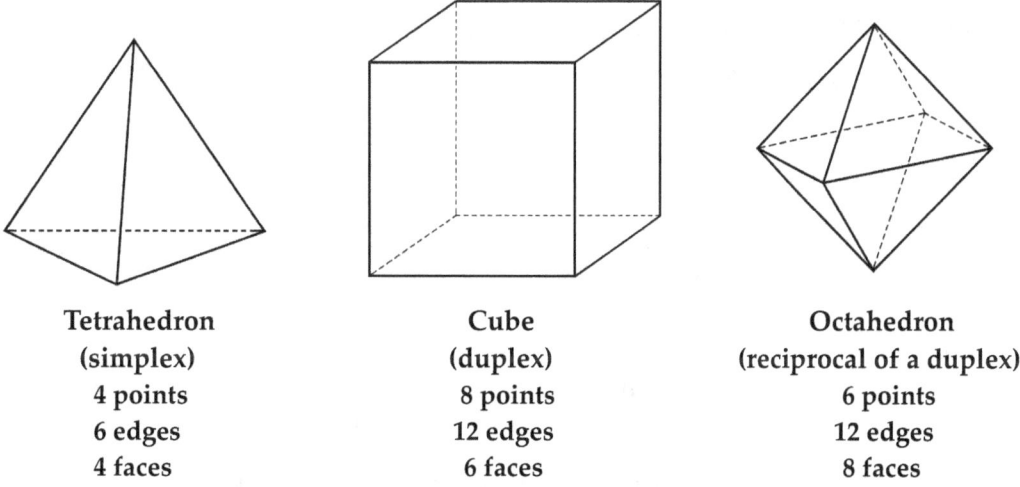

| Tetrahedron | Cube | Octahedron |
|:---:|:---:|:---:|
| (simplex) | (duplex) | (reciprocal of a duplex) |
| 4 points | 8 points | 6 points |
| 6 edges | 12 edges | 12 edges |
| 4 faces | 6 faces | 8 faces |

**Figure 1.** The pivotal 3-D elements in the 3 strands of regular convex polytopes, with their sub-elements listed.

Each regular convex polytope is described by its dimension and its sub-elements, that is, the points, segments, faces, solids, etc., that make up the total polytope. If the dimension of the polytope is **d**, it will be denoted as a **d**-polytope and it will have **d** types of sub-elements which are polytopes of dimensions 0, 1, 2, 3, . . . , **d** − 1.

The "boundary" of a convex **d**-polytope separates its interior from its exterior and is made up of **(d–1)**-polytopes. Each polytope in the boundary is made up of **(d–2)**-polytopes, forming its own boundary, and so on until we reach the **1**-polytope (a line-segment), whose boundary elements are points (**0**-polytopes). For instance, a cube (**3**-duplex) has 3 types of sub-elements, **k**-duplexes, **k** = 0, 1, and 2, each forming boundaries for the next higher dimensional duplex. The boundary elements of a cube are its faces (squares) and the boundary elements of squares are edges whose boundary elements are their endpoints.

To be a **convex d**-polytope means that all the interior points of a line-segment, joining a point on one **(d – 1)**-polytope boundary element to a point on any other **(d – 1)**-polytope boundary element, are also contained in the interior of the **d**-polytope.

To be a **regular d**-polytope means that all the **(d – 1)**-polytope sub-elements are geometrically congruent and will meet other **(d – 1)**-polytope sub-elements along a **(d – 2)**-polytope sub-element at equal inclinations. Additionally, all the sub-elements of a given dimension are congruent and will meet other sub-elements of that dimension along sub-elements of a dimension one less than the given dimension and at equal inclinations. For instance, faces will meet along line-segments, line-segments will meet at points, and these intersections will occur at equal angles, determined for each pair of sub-elements of a given type. Sub-elements of a given dimension, say k, that share a common sub-element of dimension k -1 are said to be *adjacent*. The simplest polytope that can be drawn in any dimensional space is called a **simplex**.

## 1.2  Simplexes

The simplest regular convex polytopes are point, line-segment, triangle, tetrahedron, pentatope (the simplest figure in 4-dimensions, see Fig. 3), and so on, forming a strand or sequence of geometric figures through **d**-dimensions called "**simplexes**".

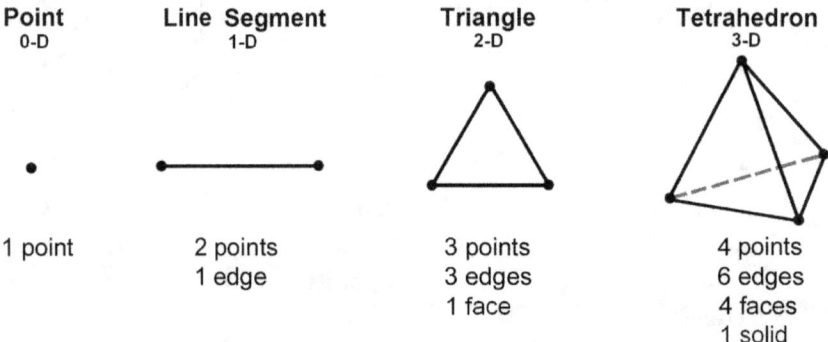

**Point**
0-D

**Line Segment**
1-D

**Triangle**
2-D

**Tetrahedron**
3-D

1 point

2 points
1 edge

3 points
3 edges
1 face

4 points
6 edges
4 faces
1 solid

**Simplexes of the dimensional spaces 0-3**
**Figure 2**

Each simplex figure in **d**-dimensions consists of d + 1 points (vertices) all of which do not lie in a **(d – 1)**-space. Each vertex is connected to the other d vertices with line-segments (edges). The number of edges contained in a **d**-simplex is equal to the number of combinations of d + 1 points taken two at a time. For instance, the number of edges of a tetrahedron (4 points) is equal to 4 choose 2, C(4, 2) = 6 (the notation C(r, c) is the binomial coefficient, r choose c.) Each simplex of a given order, or dimension, is composed of elements that are simplexes of lesser orders. The boundary of a **d**-simplex is made up of a set of d + 1 boundary simplexes of dimension **d – 1.**

For example, a tetrahedron, the **3**-simplex, is made up of points (simplexes of order 0), line segments (simplexes of order 1), triangles (simplexes of order 2), and one simplex of order 3, that is, the tetrahedron itself. The boundary of a tetrahedron (**3**-simplex) is a set of 3 + 1 = 4 equilateral triangles (**2**-simplexes).

A square is not a simplex since it consists of more than 3 defining points in a 2-dimensional space. Our intuition seems to support what we expect of the strand of simplex figures in higher dimensional spaces. A drawing of a regular and convex **pentatope**, a **4**-simplex, appears in Fig. 3. A pentatope does not exist in lower-dimensional spaces, dimensions 0, 1, 2, and 3. However, the point, the line segment, the triangle and the tetrahedron all exist as simplexes of orders, 0, 1, 2, and 3 respectively, in three and higher dimensional space.

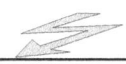

### Section 1.2

| | |
|---|---|
| **1.** | How many simplexes of order 2 are found in a tetrahedron? |
| **2.** | How many simplexes of order 1 are found in a tetrahedron? |
| **3.** | How many simplexes of order 0 are found in a tetrahedron? |
| **4.** | What is the total number of elements (sub-elements plus the tetrahedron itself) found in a tetrahedron? |

The simplest 4-dimensional polytope is a **pentatope** (a **4**-simplex). All drawings (and pictures) of objects (including simplexes) are rendered as two dimensional figures on paper. To draw a pentatope, start with a tetrahedron, then add a 5th point to the picture and connect it to each of the other points (vertices) of the tetrahedron. (Note: the 5th point is not to be considered in the original 3-D space of the tetrahedron, but all the points and lines are rendered in the 2-D space of the page to form a picture. Note also, the picture of the 4-D figure appears in a 2-D drawing as a pentagon with its diagonals.) There are C(5, 2) = 10 edges contained in a pentatope.

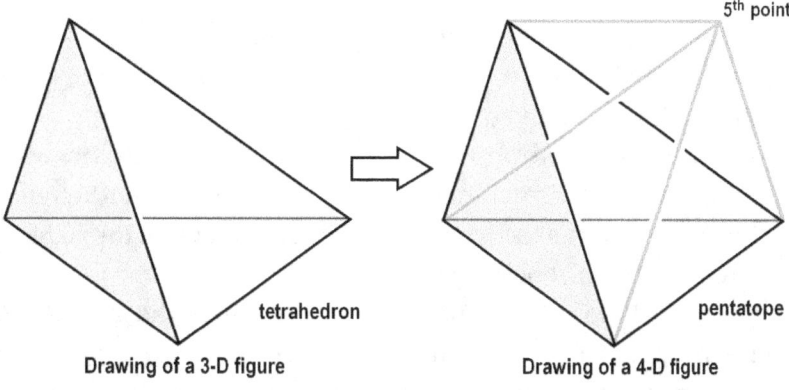

Figure 3. Transitioning from 3-D to 4-D, adding a 5th point.

Each pentatope, a simplex of order 4, is composed of elements that are simplexes of lesser orders. The pentatope contains

<u>5 points</u>
(simplexes of order 0), C(5, 1) =  5,
**A, B, C, D, E**

<u>10 edges</u>
(simplexes of order 1), C(5, 2) = 10,
**AB, AC, AD, AE,
BC, BD, BE,
CD, CE,
DE**

<u>10 triangles</u>
(simplexes of order 2), C(5, 3) = 10,
**ABC, ABD, ABE, ACD, ACE, ADE,
BCD, BCE, BDE,
CDE**

<u>5 tetrahedrons</u>
(simplexes of order 3), C(5, 4) =  5,
**I, II, III, IV, V**

The 5 tetrahedrons forming the **boundary** of the pentatope are pictured in Figure 4.  Even though it appears that the 5 tetrahedrons occupy all of the interior space of the pentatope, they do not occupy any of the interior space of the pentatope. The tetrahedrons are 3 dimensional boundary elements and, as such, they cannot enclose any part of the 4 dimensional space enclosed by the pentatope.  The same thing is true of the tetrahedron, its boundary elements are triangles and triangles cannot enclose any part of the 3 dimensional space enclosed by the tetrahedron.

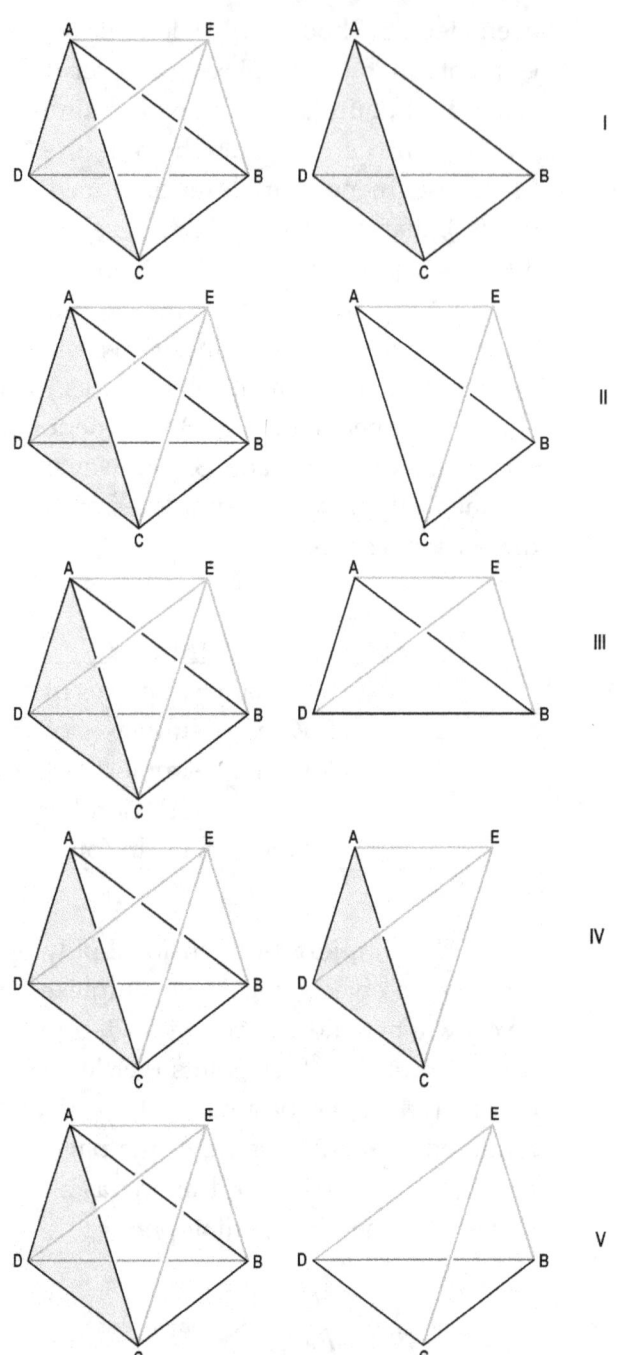

**Figure 4.  The Boundary of the Pentatope Consists of 5 Tetrahedrons
(shown above on the right, I, II, III, IV, V).**

Each of the 5 tetrahedrons shown shares each of its faces with exactly one of the other 4 tetrahedrons.  For example, tetrahedrons **I** and **II** share the face **ABC** and **I** and **IV** share the face **ACD** (the shaded face). Can you name the faces shared by **I** and **III**, and **I** and **V**?  _____  _____

In summary, the composition of each simplex of a given order, **d**,  is given in the table on the next page.

| Simplex | Subspaces | | | | |
|---|---|---|---|---|---|
| | Points (Order 0) | Line Segments (Order 1) | Triangles (Order 2) | Tetrahedrons (Order 3) | Pentatopes (Order 4) |
| Point | 1 | 0 | 0 | 0 | 0 |
| Line Segment | 2 | 1 | 0 | 0 | 0 |
| Triangle | 3 | 3 | 1 | 0 | 0 |
| Tetrahedron | 4 | 6 | 4 | 1 | 0 |
| Pentatope | ___ | ___ | ___ | ___ | 1 |

---

5.     Complete the row in the table above for the pentatope.

6.     What is the total number of elements (sub-elements plus the **4**-simplex itself) found in a **4**-simplex, the pentatope?

---

If you have recognized part of Pascal's Triangle in this table, you are right! Indeed, the number of vertices, edges, triangles, tetrahedrons, and so on, contained in each **d**-simplex is $C(d + 1, 1)$, $C(d + 1, 2)$, $C(d + 1, 3)$, $C(d + 1, 4)$, and so on, respectively (where $C(d + 1, j)$ are binomial coefficients for $j = 1$ to $d + 1$, which includes the **d**-simplex itself when $j = d + 1$.) Except for the leading 1, each row of Pascal's Triangle gives the number of sub-elements contained in each simplex. The simplex is of an order, dimension, equal to the row number in Pascal's Triangle minus one. With Pascal's Triangle as your guide, you can investigate simplexes of any order. For example, one property of Pascal's Triangle is that the sum of the elements of the r-th row is equal to $2^r$. Thus, we can conclude that the total number of sub-elements of a **r**-simplex, including the **r**-simplex itself, is equal to $2^{r + 1} - 1$ (minus 1 since the rows in the table start without the leading 1's found in Pascal's Triangle).

Another property of Pascal's Triangle is that the alternating sum of the numbers on a given row is equal to zero. For instance, $1 - 4 + 6 - 4 + 1 = 0$ and $-1 + 4 - 6 + 4 - 1 = 0$. Without the leading 1 on each row, we can conclude that

*the alternating sum of the number of subspaces of a simplex is equal to 1,* that is,

| | |
|---|---|
| **Point** | **1 vertex = 1** |
| **Line Segment** | **2 vertices – 1 edge = 1** |
| **Triangle** | **3 vertices – 3 edges + 1 triangle = 1** |
| **Tetrahedron** | **4 vertices – 6 edges + 4 faces – 1 tetrahedron = 1** |
| | **V – E + F – 1 = 1** |

The last formula is equivalent to **V – E + F = 2**, Euler's Formula for polyhedrons. We will say that **Euler's formula for simplexes** is *"the alternating sum of the number of subspaces of a simplex is equal to 1"*.

---

7.     Using the information for the pentatope in the previous table, check the formula: **V – E + F – T + 1 = 1**, where **T** represents the number of tetrahedrons we would find in a pentatope.

8.     What is the total number of elements (sub-elements plus the

**5**-simplex itself) found in a **5**-simplex?  Alternating sum?

The table below lists the simplexes for d = 0 to 5 with the number of sub-elements found in each one, see [1], pp. 292-295.  We included the original **d**-simplex as another element (the final 1 in each row of the triangular array).

| d-simplex | Number of Elements of each Type | | | | | | Total $2^{d+1}-1$ | Alternating Sum |
|---|---|---|---|---|---|---|---|---|
| | 0-D | 1-D | 2-D | 3-D | 4-D | 5-D | | |
| 0-Simplex (Point ) | 1 | | | | | | 1 | 1 |
| 1-Simplex (Line Segment) | 2 | 1 | | | | | 3 | 1 |
| 2-Simplex (Triangle) | 3 | 3 | 1 | | | | 7 | 1 |
| 3-Simplex (Tetrahedron) | 4 | 6 | 4 | 1 | | | 15 | 1 |
| 4-Simplex (Pentatope) | 5 | 10 | 10 | 5 | 1 | | 31 | 1 |
| 5-Simplex | 6 | 15 | 20 | 15 | 6 | 1 | 63 | 1 |

The triangular array (bold) in the table above is Pascal's Triangle (rows 1 through 6) with the 1's stripped away from the left-hand edge.   This triangular array, showing just the first six rows, is called the **Simplex Triangle** and we will refer to the six rows as rows 0 through 5.

---

9.    How many pentatopes form the boundary of the 5-simplex?
10.    Predict the elements of the row corresponding to **d** = 6 of the Simplex Triangle and find the total for the row as well as the alternating sum. (The entries in row 6 will be the numbers of sub-elements of the six dimensional **6**-simplex.)

---

The mathematical literature pertaining to Pascal's Triangle is extensive.  The many properties and patterns of Pascal's Triangle will apply to the Simplex Triangle, when taking into account that the only difference is the absence of the leading 1's in each row of the Simplex Triangle.   For instance, the alternating sum of the entries in each row of Pascal's Triangle  is equal to 0, but adjusting for the fact that the Simplex Triangle is without the leading 1's on each row, we find the alternating sum of the entries of the Simplex Triangle is equal to 1.   We will use this fact in our discussion of Euler's Formula in Section **2.3**.

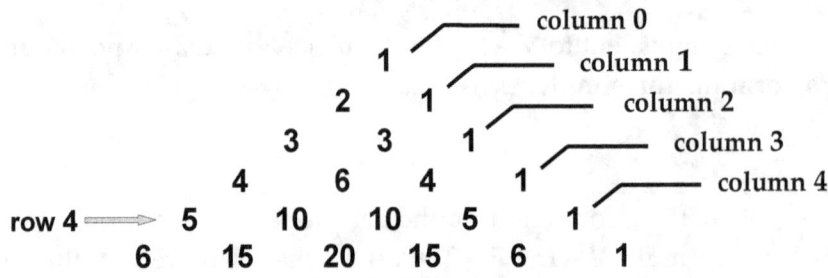

**Rows 0 through 5 of the Simplex Triangle**

## 1.3 Simplexal Numbers

The simplexal numbers are figurate numbers of higher dimensions. Sequences of these numbers are found in the columns of the Simplex Triangle (and Pascal's Triangle).

The **triangular number** sequence is found in column 1 of the Simplex Triangle (column 2 of Pascal's Triangle) and are based on the triangle simplex. The sequence is 1, 3, 6, 10, . . ..

**Triangular Numbers**

The **tetrahedral number** sequence is found in column 2 of the Simplex Triangle (column 3 of Pascal's Triangle) and are based on the tetrahedron simplex. The sequence is 1, 4, 10 , 20, . . ..

The numbers in column 3 of the Simplex Triangle (column 4 of Pascal's Triangle) are called **pentatopal numbers**. The sequence is 1, 5, 15, 35, . . ..

Triangular numbers, tetrahedral numbers, and pentatopal numbers are sequences of **simplexal** numbers of 2, 3 and 4 dimensions, respectively. Elements in these sequences correspond to extensions of simplex figures by adding on layers of the preceding lesser dimensional simplexal numbers. For example, the tetrahedral numbers (3-D) are generated by adding layers of consecutive triangular numbers (2-D) and the pentatopal numbers (4-dimensions) are generated by adding layers of consecutive tetrahedral numbers (simplexal numbers of 3-dimensions).

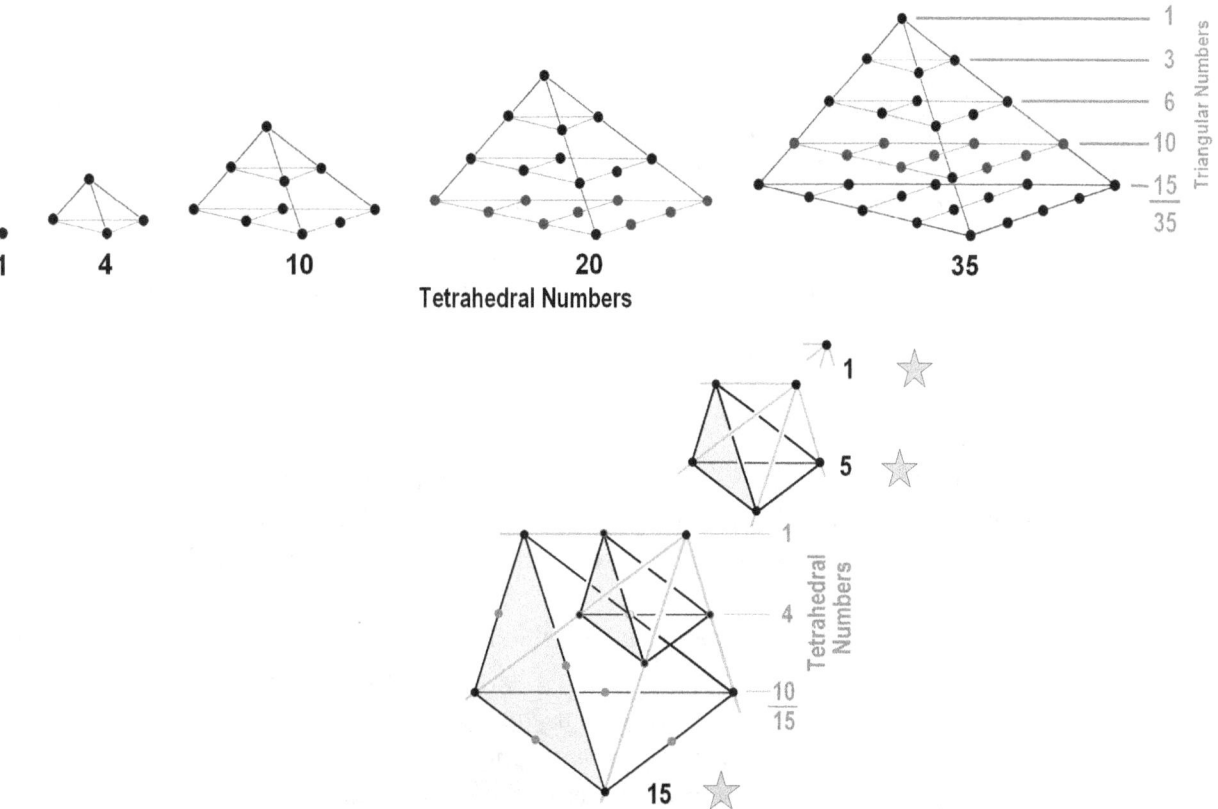

**Tetrahedral Numbers**

**The first 3 pentatopal numbers, 1, 5, 15.**

These values are also found by using the "**hockey stick**" property for Pascal's Triangle [3] (see the figure below).

**Sequences of Simplexal Numbers**

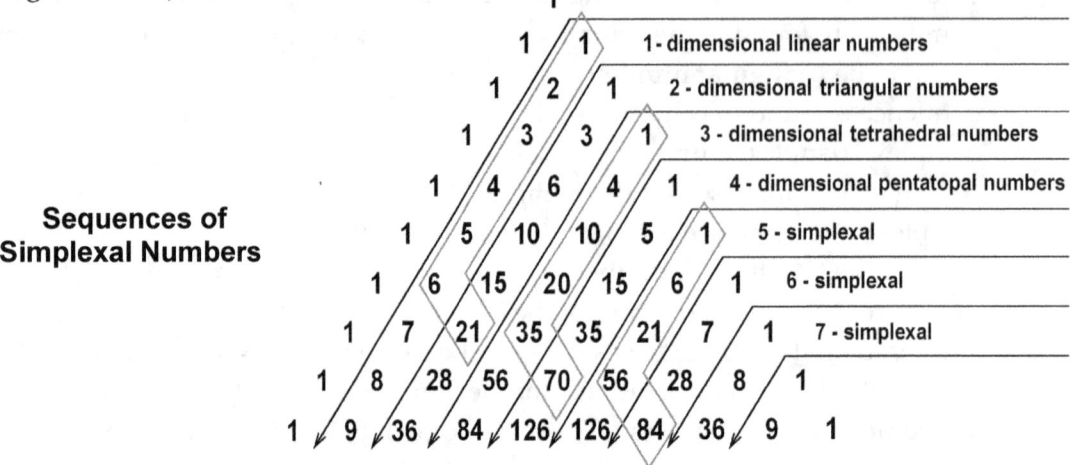

If you locate 70 in Pascal's Triangle above, you can identify it as being in row 8 and column 4, $C(8, 4) = 70$ which is the fifth pentatopal number and the sum of the first 5 tetrahedral numbers, $1 + 4 + 10 + 20 + 35 = 70$. All the simplexal numbers are found in Pascal's Triangle. In fact, Pascal's Triangle could be called a triangular array of simplexal numbers. In general,

$$C(r, c) = \text{the } (r - c + 1)\text{th  c-simplexal number.}$$

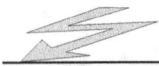 **Section 1.3**

1.    Based on Pascal's Triangle and the hockey-stick property, find
    **(a)** the 7th triangular number,  **(b)**  the 6th tetrahedral number,
    **(c)** the 6th pentatopal number,  **(d)** the 5th 5-simplexal number,
    **(e)** the 5th 6-simplexal number.

2.    The first three 7-simplexal numbers are 1, 8, and 36. Find the next one.

## 1.4  Duplexes

Elements of the second strand of regular convex polytopes are point, line-segment, square, cube, tesseract (a hypercube, a figure in 4 dimensions, see Fig. 7 in the next section), and so on, forming a strand or sequence of geometric polytopes through **d**-dimensions called "duplexes".

> *(Note: the term "duplex" is not widely used to name this strand of polytopes. Indeed, the term duplex is deeply ingrained in the English language to mean other things, so for the purposes of this text, we ask that you, the reader, temporarily suspend your understanding of the term as it is commonly used and adopt the meaning being presented here. Coxeter [1] refers to elements in this strand as "measure polytopes" and D. M. Y. Sommerville [2] refers to them as "regular orthotopes". Some authors use the term "hypercube" to refer to any higher dimensional figure in this strand.*

We will see, shortly, the term "duplex" accurately describes this strand, since each element in the strand is formed from the one immediately preceding it by a process called **geometric duplication** (see below). The term also fits nicely with the term "simplex", and like that term, duplexes refers to the entire strand of these particular polytopes.

In 2-D a duplex is a square, in 3-D it is a cube, in 4-D it is a tesseract or hypercube, and so on. We will refer to each duplex as a **d**-duplex when the specific dimension, d, of the space is needed in the discussion. For instance, a line-segment is a **1**-duplex, a square is a **2**-duplex, and so on. (Note: a **0**-duplex is a point.)

> **Define duplex:** *A regular convex **d**-polytope, or a block of d-dimensional geometric space (d ≥ 1). The block of geometric space is bounded by 2d duplexes of dimension (d − 1). A **d**-duplex, consists of $2^d$ vertices with d edges coincident at each vertex. The edges are all equal in length and orthogonal with each other at each vertex.*

## Geometric duplication

One form of drawing a duplex in the next higher dimension is a process called *geometric duplication*, as in [3] pp. 31-36. In this process, a rendering of the next higher dimensional figure is obtained by duplicating the figure of the current dimension in a displaced position, then the corresponding vertices of the two figures are connected with line-segments, completing the rendition, see Figures 5 and 6.

**The Process of Geometric Duplication**

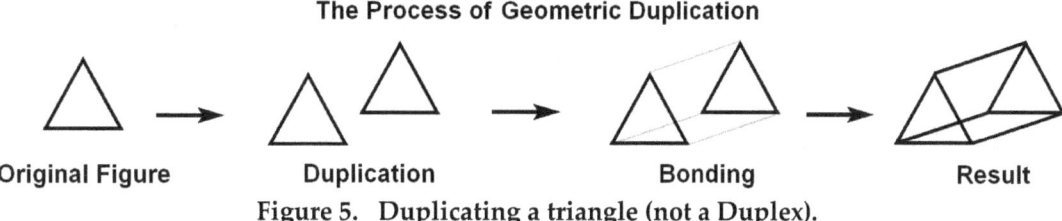

Original Figure     Duplication     Bonding     Result

**Figure 5. Duplicating a triangle (not a Duplex).**

In Fig. 5, the resulting figure could be considered a 3-dimensional prism (not a duplex). Of course, the rendition is a 2-D drawing, representing a higher dimensional figure.

We will now explore this process of geometric duplication, starting with a single point, and going through several stages to form duplexes.

**Figure 6. A point duplicates to a line-segment, a line-segment duplicates to a square, and a square duplicates to a cube.**

|  | Original | Duplication | Bonding | Result |
|---|---|---|---|---|
| **Start** |  |  |  | • |
| **Stage 1** | • | •  • | •—• | —— |
| **Stage 2** | •—• | ⌶ | ▢ | ▢ |
| **Stage 3** | ▢ | ▢▢ | cube | cube |

**Section 1.4**

1.      Count the total number of geometric sub-elements in the resulting figure of each of the stages we have developed thus far and complete the table on the next page.

|  | Result | Points | Edges | Faces | Solids |
|---|---|---|---|---|---|
| **Start** | • | 1 | | | |
| **Stage 1** | •——• | 2 | 1 | | |
| **Stage 2** | □ | 4 | 4 | 1 | |
| **Stage 3** | (cube) | 8 | —— | —— | —— |

Notice that the sequence, 1, 2, 4, 8, of points in the table of Exercise 1 above, indicates that the number of points in the resulting figure is double the preceding number. The process of geometric duplication explains why a **d**-duplex has two times as many vertices as a (**d − 1**)-duplex. It also explains why the number of edges of a **d**-duplex is two times the number of edges of the (**d − 1**)-duplex plus the number of vertices of the (**d − 1**)-duplex. For example, the number of edges of a cube is 2 times the number of edges of a square plus the number of vertices of the square. That is, $2 \cdot 4 + 4 = 12$.

2.      If we duplicate the cube and bond it, the resulting figure, a **4**-duplex, is called a **tesseract** or **hypercube**. Refer to the discussion in the preceding paragraphs, and predict the number of vertices and edges of the **4**-duplex.

From Exercise 1 of this section, we find the total number of elements at each stage are:

| | | |
|---|---|---|
| Start | 1 | = 1 |
| Stage 1 | 3 | = 2 + 1 |
| Stage 2 | 9 | = 4 + 4 + 1 |
| Stage 3 | 27 | = 8 + 12 + 6 + 1 |

Notice that the totals, 1, 3, 9, and 27, form a sequence of powers of three, $3^0$, $3^1$, $3^2$, and $3^3$.

3.      Predict the total number of elements of a tesseract at stage 4 of the process of geometric duplication.

If we write the number 3 as the binomial 2 + 1 and then expand, we find:

$$1 = (2 + 1)^0 = 1 \cdot 2^0 = \mathbf{1}$$
$$3 = (2 + 1)^1 = 1 \cdot 2^1 + 1 \cdot 2^0 = \mathbf{2 + 1}$$
$$9 = (2 + 1)^2 = 1 \cdot 2^2 + 2 \cdot 2^1 + 1 \cdot 2^0 = \mathbf{4 + 4 + 1}$$
$$27 = (2 + 1)^3 = 1 \cdot 2^3 + 3 \cdot 2^2 + 3 \cdot 2^1 + 1 \cdot 2^0 = \mathbf{8 + 12 + 6 + 1}$$

Notice, each term in these expansions (the numbers in bold in the far right-hand members) represents the number of each type of geometrical sub-element present in the resulting duplex at each stage of duplication. For instance, the numbers **4, 4,** and **1** are the number of points (vertices), line-segments (edges), faces (squares) in stage 2, forming the **2-duplex** (square) and the numbers **8, 12, 6,** and **1,** in order, are the number of points, line-segments faces, and solids (cubes) in stage 3, forming the **3-duplex** (cube).

In Chapter 2 we will demonstrate that this pattern, found in the expansions of the binomial (2 + 1), provides the terms that define the types, in order, and the numbers of sub-elements of a **d**-duplex. In higher dimensions the (**d** − 1)-duplex sub-elements forming the boundary of a **d**-duplex are sometimes called "cells".

4.      Find the 4th and 5th powers of 3.
5.      Write the binomial expansions for $(2 + 1)^4$ and $(2 + 1)^5$.
6.      Begin with the result of stage 3 (cube) and draw the result of the next stage of geometric duplication (then see Fig. 8).
7.      Complete the following table (refer to Exercises 1 and 2 and the preceding discussions):

|         | Total Elements |     | Points | | Edges | | Squares | | Cubes | | Hypercubes |
|---------|----------------|-----|--------|---|-------|---|---------|---|-------|---|------------|
| Start   | 1              | =   | 1      |   |       |   |         |   |       |   |            |
| Stage 1 | 3              | =   | 2      | + | 1     |   |         |   |       |   |            |
| Stage 2 | 9              | =   | 4      | + | 4     | + | 1       |   |       |   |            |
| Stage 3 | 27             | =   | 8      | + | 12    | + | 6       | + | 1     |   |            |
| Stage 4 | 81             | =   | ___    | + | ___   | + | ___     | + | ___   | + | ___        |

### Sub-elements of a duplex

In general, the number of *types* of sub-elements of a **d**-duplex is d.

A **d**-duplex consists of the following quantities of sub-elements:

| | |
|---|---|
| the number of vertices, **0**-duplexes, is | $C(d, 0) \cdot 2^d = 2^d$, |
| the number of edges, **1**-duplexes, is | $C(d, 1) \cdot 2^{(d-1)}$, |
| the number of faces, **2**-duplexes, is | $C(d, 2) \cdot 2^{(d-2)}$, |
| the number of cubes, **3**-duplexes, is | $C(d, 3) \cdot 2^{(d-3)}$, |
| . . . | |
| the number of (**d** − **1**)-duplexes, is | $C(d, d-1) \cdot 2^1 = 2d$, |

where $C(d, j)$ are binomial coefficients for j = 0 to d − 1. If we include the **d**-duplex in the list, it would be given by $C(d, d) \cdot 2^0 = 1$. The total number of sub-elements of a d-duplex is equal to $3^d - 1$, or if we include the **d**-duplex, the total number of elements found in a **d**-duplex is equal to $3^d$.

---

8.    Use the formula $C(d, 2) \cdot 2^{(d-2)}$ and find the number of faces contained in the **4**-duplex, the tesseract, and the **5**-duplex.

---

The table below lists the duplexes for **d** = 0 to 5 with the number of sub-elements found in each one. We included the original **d**-duplex as another element (the final 1 in each row of the triangular array).

| d-duplex | Number of Elements of each Type | | | | | | Total $3^d$ | Alternating Sum |
|---|---|---|---|---|---|---|---|---|
| | 0-D | 1-D | 2-D | 3-D | 4-D | 5-D | | |
| 0- Duplex (Point) | 1 | | | | | | 1 | 1 |
| 1- Duplex (Line-Segment) | 2 | 1 | | | | | 3 | 1 |
| 2- Duplex (Square) | 4 | 4 | 1 | | | | 9 | 1 |
| 3- Duplex (Cube) | 8 | 12 | 6 | 1 | | | 27 | 1 |
| 4- Duplex (Hypercube) | 16 | 32 | 24 | 8 | 1 | | 81 | 1 |
| 5- Duplex | 32 | 80 | 80 | 40 | 10 | 1 | 243 | 1 |

The triangular array (bold) in the preceding table, showing rows 0 through 5, is called the **Duplex Triangle**. Like the Simplex Triangle, the alternating sum of the entries in each row of Duplex Triangle is also equal to 1. We will use this fact in our discussion of Euler's Theorem in Section **2.3**.

---

9.    Predict the elements in the table above for the row corresponding to **d** = 6 of the Duplex Triangle and find the total for the row as well as ⟹

the alternating sum. (The entries in row 6 will be the numbers of sub-elements of the six dimensional **6**-duplex.)

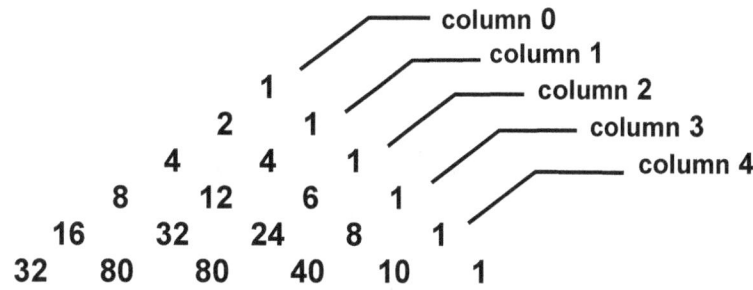

**Rows 0 through 5 of the Duplex Triangle**

## 1.5  Drawing Duplexes

The elements in Figs. 7, 8, and 9 are formed by geometric duplication, but the duplicated element is reproduced in direction different from those in our discussions thus far. Also, the figures distort angles and lengths of edges, but preserves parallelism. This has been done in order to enhance certain features of the figures that come up in the discussion to follow.

One feature we want to  emphasize in these diagrams is the number of vertex points found at each horizontal level as we scan the overall resulting figure from the top to bottom. These counts are shown just to the right of each figure. Yes, these numbers are from the rows of Pascal's Triangle and the sum of these numbers, in each case, is a power of 2, a known property of the rows of Pascal's Triangle and as expected from the definition of a duplex.

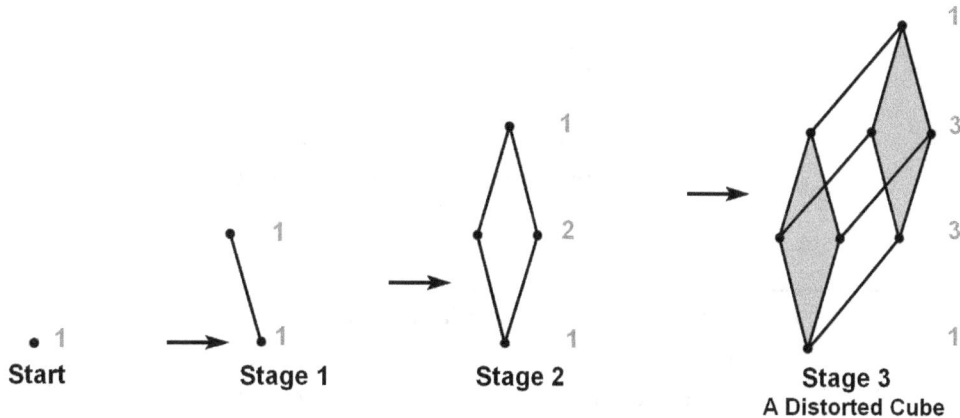

Figure 7.  First three stages of geometric duplication.

Figure 8 is a picture of Stage 4 in this series and represents a hypercube. It contains 16 points (vertices).  $1 + 4 + 6 + 4 + 1 = 16 = 2^4$.

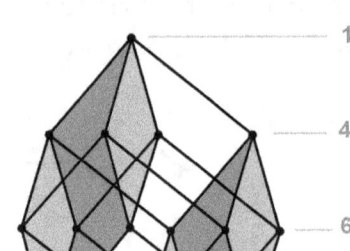

**Stage 4**

**Figure 8.   A Distorted Hypercube, a 4-Duplex.**
**(duplicating a cube)**

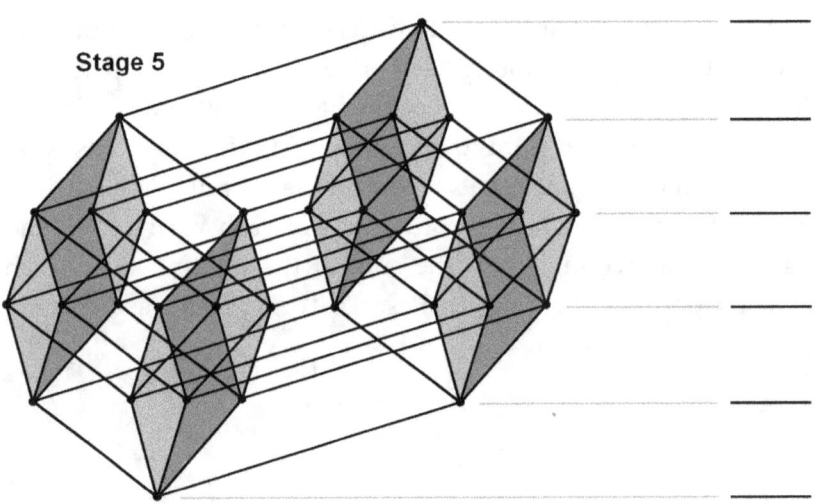

**Stage 5**

**Figure 9.    A Distorted 5-Duplex.**
**(duplicating a hypercube,  a 4-duplex)**

 **Section  1.5**

1.    Predict the number of vertices (points) for the Stage 5 duplex figure
shown in Figure 9.

2.    Extend the table from Exercise 7 of Section 1.3 to Stages 5 and 6, and
compare the totals to the numbers in Pascal's Triangle.

3.    Refer to  Fig. 10   and enter the number of points across each level.

4.    To what row of Pascal's Triangle does the set of numbers found in
Exercise 3 correspond?  What Stage of geometric duplication does
this figure represent?

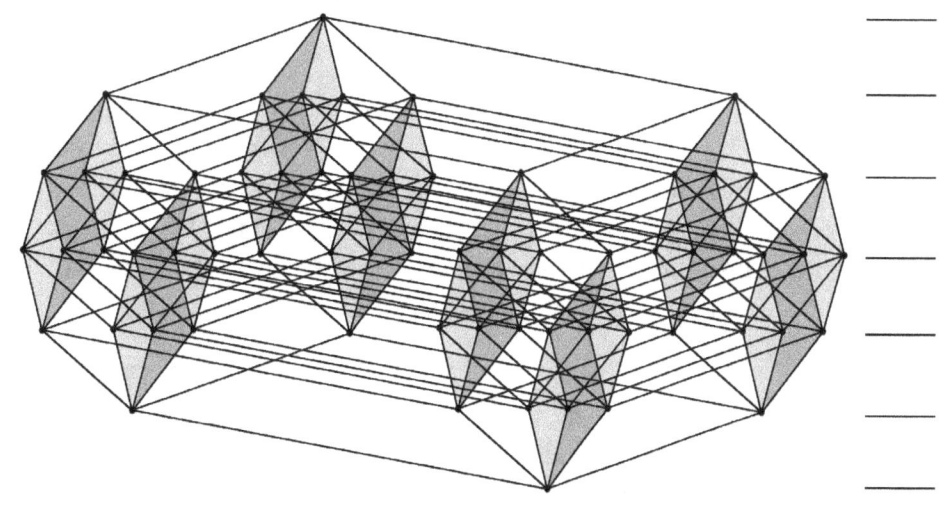

Figure 10. A Distorted 6-Duplex.

With the pictures of the cube and hypercube in Figures 7 and 8, the reader may readily count all of the other sub-elements in these figures in addition to the vertices. A way to do this is to scan the figure along each row of vertices from the top row to the bottom row. As you scan each row, from left to right, record the number of vertices (V) as we did in the preceding exercises. Then scan the row again and, at each vertex position, count the number of edges (E) below that vertex and record the total for the row. Scan the row again, this time counting and recording the number of faces (F). Repeat again, this time looking for cubes (C), and so on, up to any higher dimensional duplexes that can exist below each vertex in that row, before continuing to the next row, where you will repeat the process, row after row, until you reach the final row consisting of a single vertex.

The counts of the sub-elements of the **3**-duplex and **4**-duplex are charted in the manner just described and are shown in Figures 11 and 12.

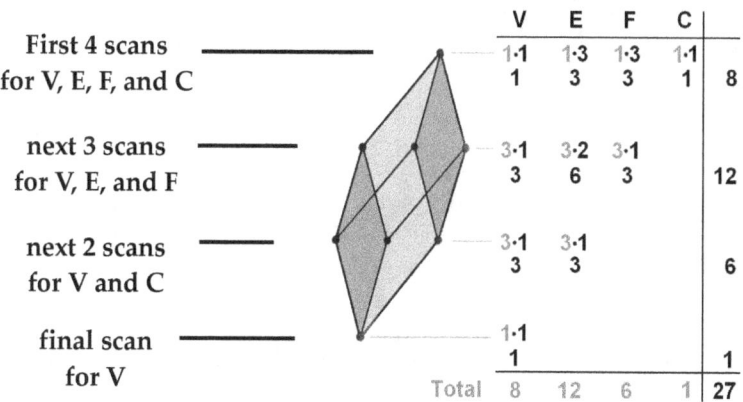

| | | V | E | F | C | |
|---|---|---|---|---|---|---|
| First 4 scans | | 1·1 | 1·3 | 1·3 | 1·1 | |
| for V, E, F, and C | | 1 | 3 | 3 | 1 | 8 |
| next 3 scans | | 3·1 | 3·2 | 3·1 | | |
| for V, E, and F | | 3 | 6 | 3 | | 12 |
| next 2 scans | | 3·1 | 3·1 | | | |
| for V and C | | 3 | 3 | | | 6 |
| final scan | | 1·1 | | | | |
| for V | | 1 | | | | 1 |
| Total | | 8 | 12 | 6 | 1 | 27 |

Figure 11. A distorted 3-duplex (cube) with its sub-elements recorded.

In Figure 12, a hypercube (H), we readily count the 16 vertices (V) and 32 edges (E). The 24 faces (squares, F) and the 8 cubes (C) can also be seen, if one scans carefully. For instance, a cube has 3 edges leading out from each of its vertices, so as you scan the row of vertices, there will be a cube below that vertex for every 3 edges leading downward from that vertex. If there are fewer than 3 edges leading downward from a given vertex, then there will be no further cubes to count in that row or subsequent rows. (Also, see Fig. 12). A hypercube has 4 edges leading out from each of its vertices and, by similar reasoning, you can count the number of hypercubes below each vertex in the row, one for every set of 4 edges leading downward from that vertex. And so on.

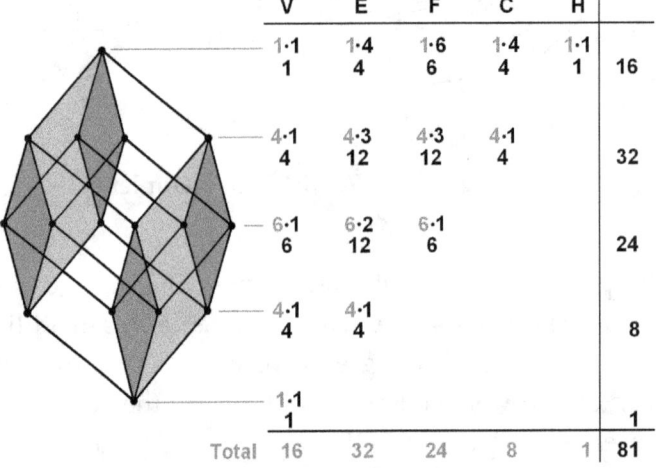

| | V | E | F | C | H | |
|---|---|---|---|---|---|---|
| | 1·1 | 1·4 | 1·6 | 1·4 | 1·1 | |
| | 1 | 4 | 6 | 4 | 1 | 16 |
| | 4·1 | 4·3 | 4·3 | 4·1 | | |
| | 4 | 12 | 12 | 4 | | 32 |
| | 6·1 | 6·2 | 6·1 | | | |
| | 6 | 12 | 6 | | | 24 |
| | 4·1 | 4·1 | | | | |
| | 4 | 4 | | | | 8 |
| | 1·1 | | | | | |
| | 1 | | | | | 1 |
| Total | 16 | 32 | 24 | 8 | 1 | 81 |

**Figure 12.   A distorted 4-duplex (tesseract) indicating its 16 vertices, 32 edges, 24 faces, and 8 cubes. The 4-duplex is formed by duplicating a 3-duplex (shaded), then connecting corresponding vertices (bonding).**

The 8 cubes recorded in Fig. 12, forming the boundary of the hypercube, are hard to visually identify, since they appear to intersect each other when viewed as a 2-D rendition. In Fig. 13 the 8 cubes are separated out for identification.

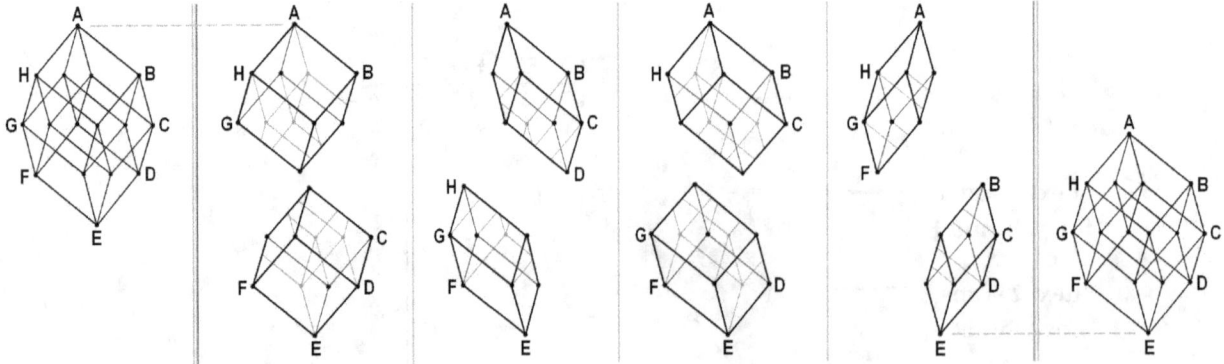

**Figure 13.   The eight cubic (distorted) sub-elements forming the boundary of the hypercube.**

The diagrams on the extreme left and extreme right of Fig. 13 are copies of the 2-dimensional representation of a distorted hypercube. However, because of the distortion, we can isolate 4 pairs of distorted cubes which are shown in the center of the figure. Half of the vertices of the hypercube are labeled to help you identify each cube and they are lined up to help you locate their positions in the hypercubes shown on the right and the left ends of the figure.

For **d**-duplexes, **d** > 4 , the sub-elements become increasingly difficult to spot in 2-D drawings. There is, however, a pattern in the counting process as shown in Figures 11 and 12.

## Enter Pascal's Pyramid

Pascal's Pyramid is a 3 dimensional array of the coefficients generated in the expansion of the trinomial $(a + b + c)^d$, with each level of the Pyramid corresponding to different values of d. If $a = b = c = 1$, then the sum of the coefficients on each level equals $3^d$, which is in agreement with our earlier result for the total number of all sub-elements of a **d**-duplex along with the **d**-duplex itself.

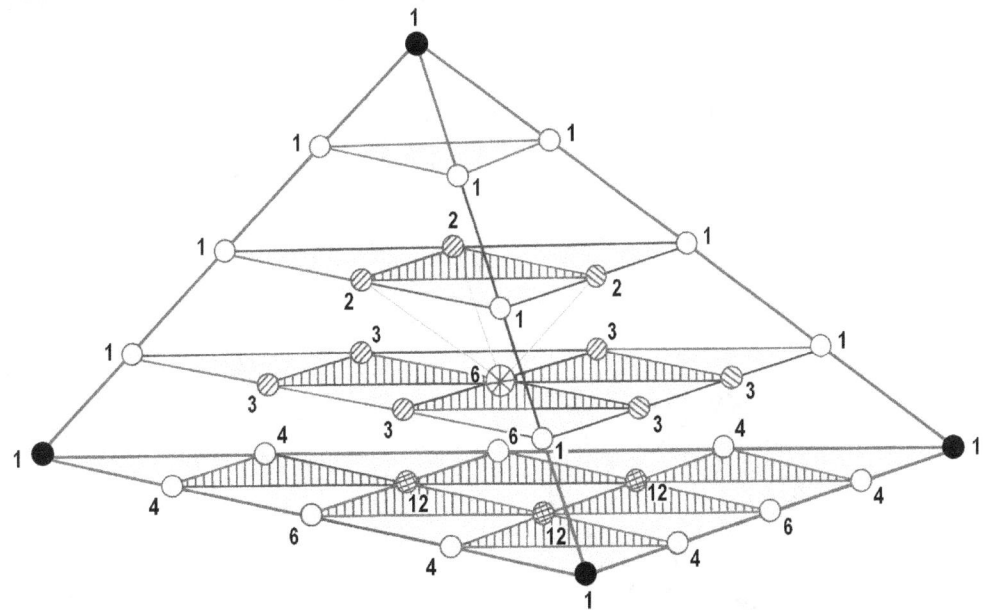

**Figure 14. Pascal's Pyramid (to level 4)**

The triangular arrays of products shown in Figures 11 and 12 (bold entries) are also the arrays of numbers found on the 3rd and 4$^{th}$ levels of Pascal's Pyramid. The array of numbers found in Figure 12 is shown below:

| | | | | | | | | | | |
|---|---|---|---|---|---|---|---|---|---|---|
| **1** | **4** | **6** | **4** | **1** | | 1 | 4 | 6 | 4 | 1 |
| **4** | **12** | **12** | **4** | | | | 4 | 12 | 12 | 4 |
| **6** | **12** | **6** | | | | | | 6 | 12 | 6 |
| **4** | **4** | | | | | | | | 4 | 4 |
| **1** | | | | | | | | | | 1 |

**Two versions of the triangular array corresponding to the counts of the sub-elements of the 4-duplex. The array is the same as the 4$^{th}$ level of Pascal's Pyramid.**

These arrays (Figs. 11 and 12) also show us an interesting pattern involving the first 3 and 4 rows, respectively, of Pascal's Triangle. For instance, each row of the triangular array corresponding to the 4th level of Pascal's Pyramid can be expressed as products of the elements of the 4th row of Pascal's Triangle with all of the elements of prior rows of Pascal's Triangle as follows: (The operations shown below are called scalar products.)

$$1^{st} \text{ row:} \quad 1 \bullet (1, 4, 6, 4, 1) = (1, \ 4, \ 6, \ 4, \ 1)$$
$$2^{nd} \text{ row:} \quad 4 \bullet (1, 3, 3, 1) \quad = (4, 12, 12, \ 4)$$
$$3^{rd} \text{ row:} \quad 6 \bullet (1, 2, 1) \quad = (6, 12, \ 6)$$
$$4^{th} \text{ row:} \quad 4 \bullet (1, 1) \quad = (4, \ 4)$$
$$5^{th} \text{ row:} \quad 1 \bullet (1) \quad = (1)$$

For a **4**-duplex (hypercube, H), we have 16 vertices (V), 32 edges (E), 24 faces (squares, F) and the 8 cubes (C).

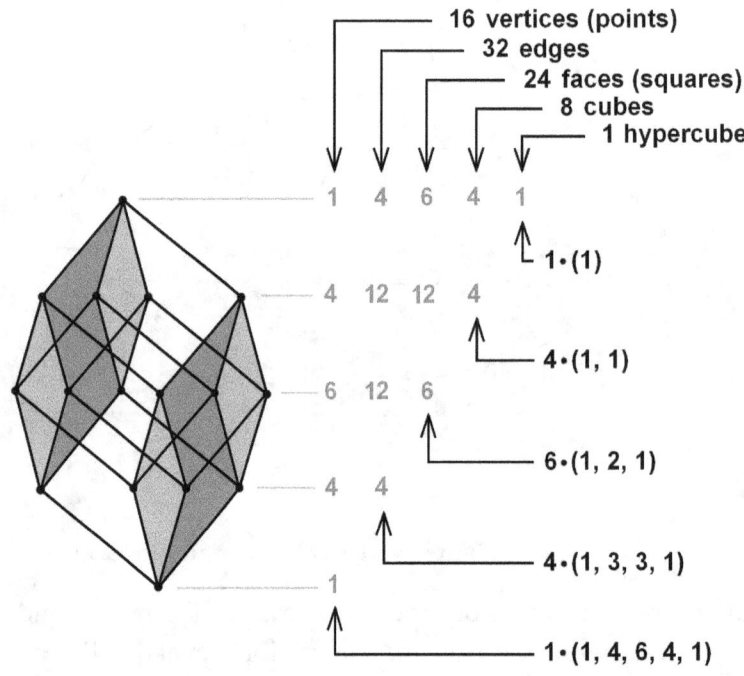

---

5.   Find the following scalar products corresponding to the 3rd level of Pascal's Pyramid.

$$1 \bullet (1, 3, 3, 1) =$$
$$3 \bullet (1, 2, 1) =$$
$$3 \bullet (1, 1) =$$
$$1 \bullet (1) =$$

6.   Fill in the table on the next page indicating the numbers of sub-elements of a cube:

| V | E | F | C | Total |
|---|---|---|---|---|
| 1 | ___ | ___ | ___ | ___ |
| 3 | ___ | ___ | | ___ |
| 3 | ___ | | | ___ |
| 1 | | | | ___ |
| 8 | ___ | ___ | ___ | 27 |

Compare your table entries to the scalar products of Exercise 5.

7.     Predict the entries in the triangular arrays corresponding to the 5th and 6th levels of Pascal's Pyramid.

8.     Find the total of all the entries in the in the triangular array corresponding to the 5th level of Pascal's Pyramid.   Same for the 6th level of Pascal's Pyramid.

9.     Express the totals found in Exercise 8 as powers of three.

## 1.6   The Diagonals of Duplexes

A **d**-duplex has $2^d$ vertices and $2^{d-1}$ diagonals.  Each diagonal consists of two vertices that are opposite of each other in the duplex.  For the   **1**-duplex (line-segment),   **2**-duplex (square) and   **3**-duplex (cube), these are easy to see and their lengths are easily found using the Pythagorean Theorem.  All the edges in the duplexes pictured below are equal to 1 unit.

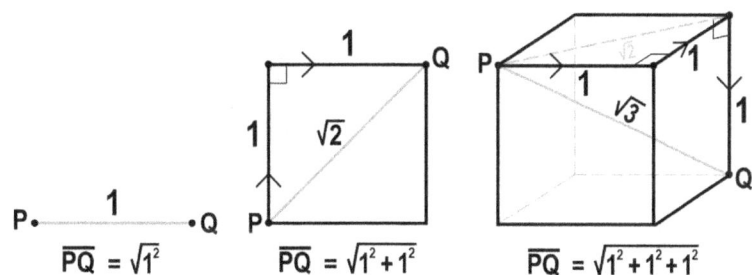

$$\overline{PQ} = \sqrt{1^2} \qquad \overline{PQ} = \sqrt{1^2 + 1^2} \qquad \overline{PQ} = \sqrt{1^2 + 1^2 + 1^2}$$

For the **4**-duplex (hypercube), finding the diagonals (there are 8 of them) and finding their lengths can be easily calculated using the extended Pythagorean Theorem.  The figure on the next page, on the left, shows a **4**-duplex (hypercube), with the peripheral vertices labeled from A to H.  A and E are one pair of opposite vertices, determining one diagonal. There are 8 boundary cubes (sub-elements) in the hypercube and in the figure on the next page,  on the right, one cube has been highlighted with the peripheral verticies ABJLGH.  To draw the diagonal, in this case from A to E, we chose one of the paths along the edges of the selected cube,  starting at vertex A and going to the end of the diagonal of the chosen cube, from A to L in this case.   The path then goes from L to E,  in a direction orthogonal to each of the edges of the cube that are coincident at L.  This segment of the path will be in the 4$^{th}$ dimension relative to the chosen cube.

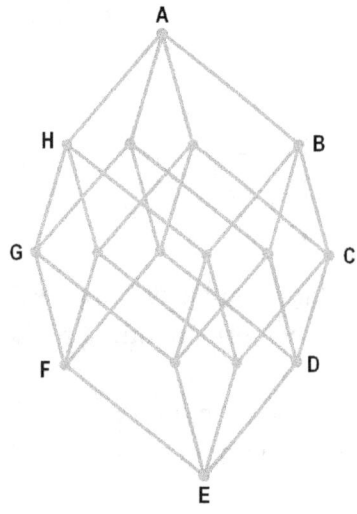 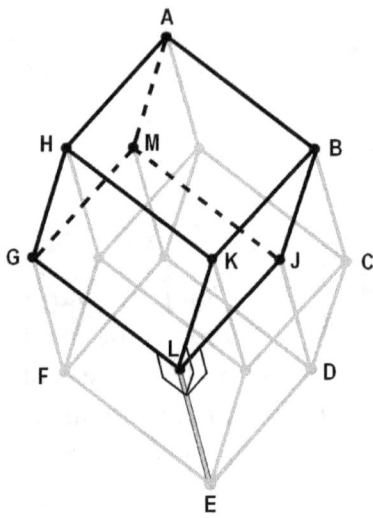

See the figure below.

The length of the diagonal, AE,  is equal to the square root of the sum of the squares of the lengths of the edges along the path ABJLE. In this case, that is

$$\sqrt{1^2 + 1^2 + 1^2 + 1^2} = 2.$$

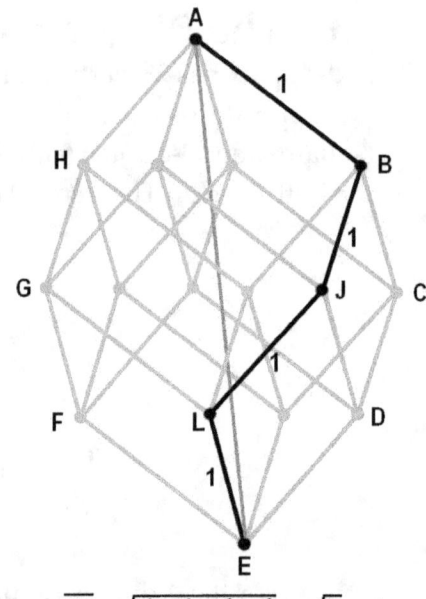

$$\overline{AE} = \sqrt{1^2 + 1^2 + 1^2 + 1^2} = \sqrt{4} = 2$$

**(Diagonal of a hypercube, 4-duplex)**

In general,

*the length of the diagonal of a **d**-duplex is equal to $\sqrt{d}$ times the length of an edge, **n**, of the duplex.*

$$\overline{diagonal} = \sqrt{n^2 + n^2 + n^2 + \ldots + n^2} = \sqrt{d \cdot n^2} = n\sqrt{d}$$

Pictured below is the "spiral" of the lengths of the diagonals of unit **d**-duplexes for **d** = 1 to 17 and **n** = 1.

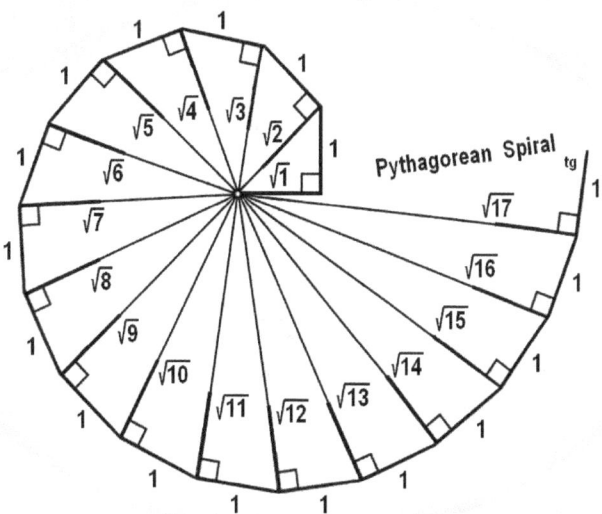

          **Section 1.6**

1.      How many diagonals do each of the following **d**-duplexes have:
        **(a)** a **3**-duplex (cube), **(b)** a **4**-duplex (hypercube), **(c)** a **5**-duplex ?
2.      Find the length of the diagonal of a cube whose edge measures 3 units.
3.      Find the length of the diagonal of a hypercube whose edge measures 3 units.
4.      Find the length of the diagonal of **d**-duplex whose edges measures 1 unit, each **(a)** for **d** = 5, and **(b)** for **d** = 9.
5.      What is the dimension of the duplex whose diagonal measures 12 units and **(a)** whose edges measures 6 units? **(b)** whose edges measures 3 units?

We also have the following <u>theorem</u> for **d**-duplexes.

**Thm.** *The sum of the squares of the lengths of all the edges of a d-duplex is equal to the sum of the squares of the lengths of all of its diagonals.*

Example:  A **3**-duplex, cube, has 4 diagonals, each of length $n\sqrt{3}$.
        The square of $n\sqrt{3}$ is equal to $3n^2$. The sum of the squares of the lengths of all the diagonals is equal to $4(3n^2) = 12n^2$.
        The cube has 12 edges, each of length **n**. The sum of the squares of the lengths of all the edges is equal to $12n^2$. The results are the same.

The proof of the theorem is as follows:

The number of edges of a **d**-duplex is equal to $C(d, 1) 2^{d-1} = d(2^{d-1})$ and the length of each edge is **n**. The sum of the squares of all of the edges is equal to $d(2^{d-1}) n^2$.

The number of diagonals of a **d**-duplex is equal to $2^{d-1}$ and the length of each diagonal is $n \sqrt{d}$. The sum of the squares of all of the diagonals is equal to $(2^{d-1}) d n^2$. The sums are equal.                                                                      ∎

In the figures below we have a visual demonstration of the theorem for the **2-duplex**, a square. Figure A starts with the square. Figure B shows the squares of the edges of the original square. Figure C shows the square of one diagonal of the original square and Figure D shows the square of the other diagonal.

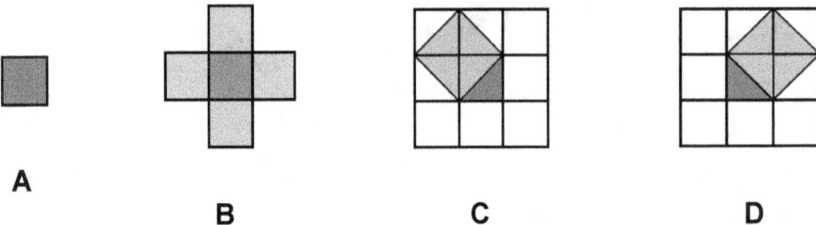

**A**

**B**                          **C**                          **D**

Figures E and F, below, show that the four parts of each of the squares of the diagonals are equal to the four parts of two of the squares on the edges.

**Note:** In general, the theorem on the previous page is also true for rectangles and boxes

**E**                          **F**

with unequal edges. That is, the sum of the of the squares of the lengths of all the edges of a box or rectangle is equal to the sum of the squares of the lengths of all of its diagonals. The box can also go through geometric duplication into 4 dimensions and the theorem still holds.

6. Given a square with edges equal to 3 units each, show that the sum of the squares of the lengths of all the edges is equal to the sum of the squares of the lengths of its diagonals.

7. Given a **4**-duplex, hypercube, with edges equal to 2 units each, show that the sum of the squares of the lengths of all the edges is equal to the sum of the squares of the lengths of all of its diagonals.

8. Given a 2 by 3 by 4 box, show that the sum of the squares of the lengths all 12 of its edges is equal to the sum of the squares of the lengths of all of its diagonals.

If we consider a unit **d**-duplex (when the edges are all equal to 1 unit), then we have the following corollary to the theorem on page 25.                                          ⟹

Corollary: *Given a unit **d**-duplex, the sum of the squares of the lengths of all of its diagonals is numerically equal to the number of its edges.*

## 1.7   The Third Strand of Regular Convex Polytopes

There is a third strand of regular convex polytopes which is the strand of **reciprocal duplexes**. For instance, consider the **3**-duplex. The **3**-duplex (cube) consists of 8 vertices, 12 edges, and 6 faces. The reciprocal of this consists of 6 vertices, 12 edges, and 8 faces. These numbers are in reverse order of the numbers of sub-elements just given for a cube, and defines an octahedron. The boundary of the octahedron consists of the 8 faces (triangles). The octahedron is the pivotal 3-D element in this, the third strand of regular convex polytopes (refer back to Fig. 1). (Note: In the world of polytopes, the strand of simplexes is its own strand of reciprocals.)

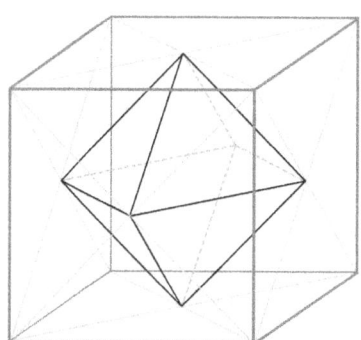

The the centers of the 6 faces of the cube in Fig. 15 are the 6 vertices of the octahedron, demonstrating the interchange of faces for vertices in these reciprocal polytopes. Indeed, if the entries in the table on page 16, Section 1.4, for the number of sub-elements of a **d**-duplex are reversed while leaving the final entry of '1' on each row, we would have the number of sub-elements of the regular convex polytopes that are the reciprocals of the duplexes. Thus, in 4-D the numbers of the sub-elements of the hyper-octahedron (also called a 16-cell) are 8 vertices, 24 edges, 32 faces (triangles), and 16 cells (octahedrons).

**Figure 15. An Octahedron Inscribed in a Cube.**

The table below lists the reciprocal duplexes for **d** = 0 to 5 with the number of sub-elements found in each one. We included the original reciprocal **d**-duplex as another element (the final 1 in each row of the triangular array).

| reciprocal d-duplex | Number of Elements of each Type | | | | | | Total $3^d$ | Alternating Sum |
|---|---|---|---|---|---|---|---|---|
| | 0-D | 1-D | 2-D | 3-D | 4-D | 5-D | | |
| 0- D (Point) | 1 | | | | | | 1 | 1 |
| 1- D (Line-Segment) | 2 | 1 | | | | | 3 | 1 |
| 2- D (Square) | 4 | 4 | 1 | | | | 9 | 1 |
| 3- D (Octahedron) | 6 | 12 | 8 | 1 | | | 27 | 1 |
| 4- D (Hyper-octahedron) | 8 | 24 | 32 | 16 | 1 | | 81 | 1 |
| 5- D (32-Cell) | 10 | 40 | 80 | 80 | 32 | 1 | 243 | 1 |

In general, the reciprocal of a **d**-duplex is a regular convex **d**-polytope with d types of sub-elements consisting of the following counts of its sub-elements:

the number of vertices (**0**-D) is equal to the number of **(d – 1)**-duplexes
the number of edges (**1**-D) is equal to the number of **(d – 2)**-duplexes
the number of faces (**2**-D) is equal to the number of **(d – 3)**-duplexes
:
:

the number of **(d – 2)**-D elements is equal to the number of **1**-duplexes (edges)
the number of **(d – 1)**-D elements is equal to the number of **0**-duplexes (vertices)

**Note:** The counts of the sub-elements of a reciprocal **d**-duplex are just the counts for the sub-elements of the **d**-duplex of the same dimension in reverse order. Also, the totals and alternating sums are the same as they are for duplexes. The properties of the triangular array of numbers for the sequence of hyper-octahedrons would parallel the properties of the Duplex Triangle.

The number of **diagonals** of the octahedron is 3 and their lengths are equal to the length of the edge of the circumscribing cube. In general, the number of diagonals of a reciprocal **d**-duplex (hyper-octahedron) is equal to **d**.

## Section 1.7

1.    Use Fig. 15 to verify that the octahedron and the cube each have 12 edges.
2.    Draw a square and join the midpoints of the sides. What figure is formed ? (The sequence of the strand of the duplexes and the strand of the reciprocals of the duplexes, each begin with the figures point, line-segment, and square. After that, the two sequences continue with the cube and the octahedron, respectively.)
3.    From the preceding table the number of faces of the 4-D Hyper-octahedron is 32. The reciprocal polytope is a **4**-duplex (hypercube) and the **4**-duplex has 32 sub-elements of what type?
4.    Verify that the alternating sum of the entries in each row of the table of elements for the reciprocal **d**-duplexes is equal to 1.
5.    Verify that the number of edges of a reciprocal **d**-duplex is equal to 4 times the (d – 1)th triangular number for d = 2, 3, 4, and 5.
6.    Verify that the number of faces of a reciprocal **d**-duplex is equal to 8 times the (d – 1)th tetrahedral number for d = 3, 4, and 5.
7.    How many diagonals exist in a 4-D Hyper-octahedron?
8.    The length of a diagonal of an octahedron is n. Show that the sum of the squares of the lengths of all of the diagonals is equal to one-half the sum of the squares of all of the edges of the octahedron. Hint: refer to Figure 15 and do some calculations.

# 2

# Developing the Duplex Triangle

## 2.1 Pascal's Triangle vs. the Duplex Triangle

In Chapter 1 we defined a geometric duplex and developed an intuitive understanding of the Duplex Triangle based on the process of geometric duplication. In this chapter our goal is to develop the Duplex Triangle in a somewhat more formal manner.

If you ask someone how many faces, edges, and vertices a cube has, I don't think the first thing they would think about would be Pascal's Triangle. However, the $3^{rd}$ row of Pascal's Triangle, without the leading 1, that is, (3  3  1), provides the basis for the answers, which are revealed when we multiply each member by the powers of two, $2^1$, $2^2$, and $2^3$, respectively, in the order faces, edges, and vertices.

$$(2^1 \bullet \mathbf{3}, \quad 2^2 \bullet \mathbf{3}, \quad 2^3 \bullet \mathbf{1}) = (6, 12, 8)$$

Actually, the Simplex Triangle is Pascal's Triangle without the leading 1 in each row, and the string (3, 3, 1) is the $3^{rd}$ row of the Simplex Triangle.

And the best part is, we will demonstrate that this type of calculation works for each higher dimensional duplex. For example, the number of cells, faces, edges, and vertices of a hypercube (a **4**-duplex) can quickly be found by multiplying the $4^{th}$ row of the Simplex Triangle by the powers of two, $2^1$, $2^2$, $2^3$, and $2^4$, respectively.

$$( 2^1 \bullet \mathbf{4}, \quad 2^2 \bullet \mathbf{6}, \quad 2^3 \bullet \mathbf{4}, \quad 2^4 \bullet \mathbf{1}) = (2 \bullet \mathbf{4}, \quad 4 \bullet \mathbf{6}, \quad 8 \bullet \mathbf{4}, \quad 16 \bullet \mathbf{1}) = (8, 24, 32, 16)$$

This tells us the hypercube consists of 8 cubes (cells), 24 faces, 32 edges, and 16 vertices.

If we reverse the order of the elements of (8, 24, 32, 16), the result is (16, 32, 24, 8) and these are the first 4 elements of the $4^{th}$ row of the Duplex Triangle. The rows of the Duplex Triangle can be derived from the rows of the Simplex Triangle when the elements in the rows of the Simplex Triangle are reversed and multiplied by powers of two in descending order. The right-hand side of the equation below shows the factors of the first 4 elements in the $4^{th}$ row of the Duplex Triangle in terms of the $4^{th}$ row of Pascal's Triangle and the powers of 2.

$$(16, 32, 24, 8) = (16 \bullet \mathbf{1}, \quad 8 \bullet \mathbf{4}, \quad 4 \bullet \mathbf{6}, \quad 2 \bullet \mathbf{4}) = ( 2^4 \bullet \mathbf{1}, \quad 2^3 \bullet \mathbf{4}, \quad 2^2 \bullet \mathbf{6}, \quad 2^1 \bullet \mathbf{4})$$

We will express the string (16, 32, 24, 8) as the **product of two strings** as follows:

$$(2^4, 2^3, 2^2, 2^1)(\mathbf{1, 4, 6, 4}) = ( 2^4 \bullet \mathbf{1}, \quad 2^3 \bullet \mathbf{4}, \quad 2^2 \bullet \mathbf{6}, \quad 2^1 \bullet \mathbf{4}) = (16, 32, 24, 8),$$

where, in this case, the first string consists of powers of 2 and the second string consists of the entries in the 4th row of Pascal's Triangle, without the final 1.

When we include the final 1 in the string of entries in the 4th row of Pascal's Triangle and append $2^0$ to the string of powers of 2, we get the **string product** representing the 4th row of the Duplex Triangle, as follows:

$$(2^4, 2^3, 2^2, 2^1, 2^0)(1, 4, 6, 4, 1) = (2^4 \bullet 1, 2^3 \bullet 4, 2^2 \bullet 6, 2^1 \bullet 4, 2^0 \bullet 1) = (16, 32, 24, 8, 1).$$

The final 1 in the resulting string represents the **4**-duplex (hypercube). The string (16, 32, 24, 8, 1) represents the number of sub-elements of the 4-duplex in the order 16 **0**-duplexes, 32 **1**-duplexes, 24 **2**-duplexes, 8 **3**-duplexes, and 1 **4**-duplex, or, in other words, 16 vertices, 32 edges, 24 faces, 8 cubes, and 1 hypercube.

In general, the demonstration will show that

*the nth row of the Duplex Triangle is the product of the string consisting of the powers of two (from the nth power down to the zeroth power) and the string consisting of the elements of the nth row of Pascal's Triangle.*

**Section 2.1**

1.    Write down the first 4 rows of the Duplex Triangle and then rewrite those rows in factored form in terms of the first 4 rows of Pascal's Triangle and the associated powers of two. Hint: Review Section 1.3.

2.    Compute the product of the strings consisting of the 5th row of Pascal's Triangle, and the powers of 2 from the 5th power down to the zeroth power. What row of the Duplex Triangle is the result?

3.    Use the product of strings to find the sub-elements of a **6**-duplex and the elements of the 6th row of the Duplex Triangle.

## 2.2  The Demonstration

The measure of the capacities or content of duplexes in d-dimensions, d > 0, is given by the sequence of powers of n,

$$n, n^2, n^3, \ldots, n^d, \ldots$$

where n represents the length of each of the d edges coincident at a single vertex of the duplex.

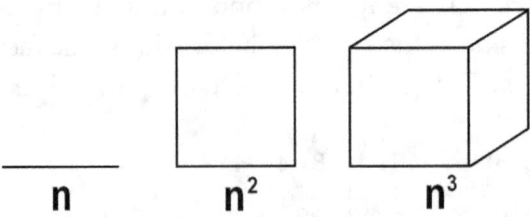

$$n \qquad n^2 \qquad n^3$$

The content, or capacity, of a **d**-duplex is $n^d$. The method of the demonstration is to cover the **d**-duplex with a shell, or hull, one unit thick. As a result, the measure of each edge is increased by 2 units, one unit on each end of each edge. The measure of the capacity of the resulting duplex is $(n + 2)^d$. From this larger **d**-duplex we remove the original interior **d**-duplex (see Figures 1 & 2). This leads to equation (1), giving us just the measure of the capacity of the covering shell, or hull:

$$(n + 2)^d - n^d = C(d, 1) \cdot 2 \cdot n^{(d-1)} + C(d, 2) \cdot 2^2 \cdot n^{(d-2)}$$
$$+ C(d, 3) \cdot 2^3 \cdot n^{(d-3)}$$
$$+ \ldots + C(d, d-1) \cdot 2^{(d-1)} \cdot n^{(d-(d-1))}$$
$$+ C(d, d) \cdot 2^d \cdot n^{(d-d)}, \tag{1}$$

or, equivalently,

$$\mathbf{(n + 2)^d - n^d = \sum C(d, j) \cdot 2^j \cdot n^{(d-j)}}, \text{ for } j = 1 \text{ to } d, \tag{2}$$

where $C(d, j)$ are the binomial coefficients for $j = 1$ to $d$, which are equivalent to the entries in row d of the Simplex Triangle.

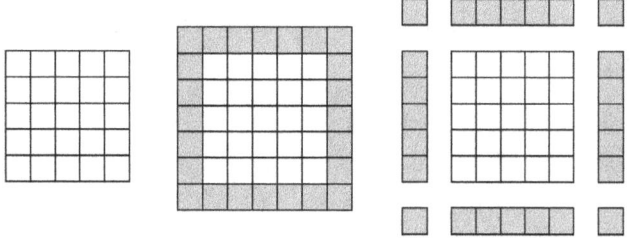

**Figure 1.** A **2**-duplex (square), border added, break-out showing
sub-elements, 4 corners and 4 edges.

The shaded parts of the figure correspond to the 4 corners and 4 edges of the original square.

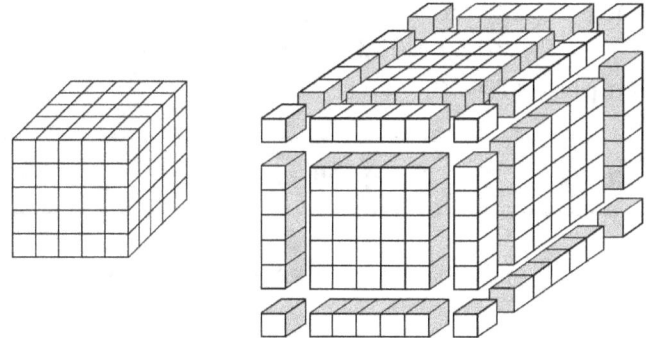

**Figure 2.** A **3**-duplex (cube) and a covering shell or hull.
Showing only the front, top, and side of the covering and
a break-out of those sub-elements.

More importantly, coded into (1) and (2) is all of the information we need to identify the number and type of each sub-element. Each term of the right member of (1) indicates three significant pieces of information.  First, the factor $n^{(d-j)}$ in the general term of (2), that is,

$$C(d, j) \cdot 2^j \cdot n^{(d-j)},$$

indicates both the type and capacity of each sub-element, specifically, the $(d-j)$-duplexes. The other factor, $C(d, j) \cdot 2^j$, gives us the number of sub-elements of the indicated type.  To illustrate equation (1) for $n = 3$ (cube), we have

$$(n + 2)^3 - n^3 = 3 \cdot 2 \cdot n^2 + 3 \cdot 2^2 \cdot n^1 + 1 \cdot 2^3 \cdot n^0$$
$$= 6 \cdot n^2 + 12 \cdot n^1 + 8 \cdot n^0. \tag{3}$$

Reading the right-hand member of (3), term by term, we conclude there are 6 faces ($n^2$), 12 edges ($n^1$), and 8 vertices ($n^0$).

In general, the number of types of sub-elements of a **d**-duplex is d.  Those types being  **0**-duplexes, **1**-duplexes, **2**-duplexes, . . . , **(d – 1)**-duplexes, also known as points (vertices), line-segments (edges), squares (faces), cubes (cells) and higher dimensional elements for $d > 3$.

The terms on the right-hand side of equation (1) enumerate the  following quantities of sub-elements of a **d**-duplex:

the number of vertices, **0**-duplexes, is    $C(d, d) \cdot 2^d = 2^d$,
the number of edges, **1**-duplexes, is    $C(d, d-1) \cdot 2^{(d-1)}$,
the number of faces, **2**-duplexes, is    $C(d, d-2) \cdot 2^{(d-2)}$,
the number of cubes, **3**-duplexes, is    $C(d, d-3) \cdot 2^{(d-3)}$,

. . .

the number of **(d - 1)**-duplexes, is    $C(d, 1) \cdot 2^1 = 2d,$

Note: The coefficients $C(d, j)$ are lined up in a order different from that presented in Chap. 1, but these expressions are equivalent to those because $C(d, j) = C(d, d-j)$ due to the symmetry of the entries in each row of Pascal's Triangle.

Let $k = d - j$ in (2) and let $N(d, k)$ represent the number of  sub-elements of the type **k**-duplexes in a **d**-duplex, then

$$\boxed{N(d, k) = C(d, d-k) \cdot 2^{d-k} = C(d, k) \cdot 2^{d-k}.} \tag{4}$$

The total number of  sub-elements of a  cube ($d = 3$) is  given by

$$N(3, 2) + N(3, 1) + N(3, 0) = 6 + 12 + 8$$
$$= 26$$
$$= 3^3 - 1.$$

**Section  2.2**

**1.**    Use the formula (4) for number of  sub-elements of the type

**k**-duplexes in a **d**-duplex, N(d, k), to find the number of faces (**2**-duplexes) of a **4**-duplex (tesseract).

2.    Use the formula, N(d, k), to find the number of cells (cubes) of a **4**-duplex (tesseract).

3.    Find the total number of sub-elements of a hypercube (tesseract) by using the following formula:

$$N(4, 3) + N(4, 2) + N(4, 1) + N(4, 0) =$$

## 2.3  The Sum and Alternating Sum of the Sub-elements of a d-Duplex

### Total number of sub-elements of a duplex

The number of sub-elements of each **d**-duplex does not depend on the length of each edge of the duplex. If we let n = 1, we have a unit duplex and (1) becomes

$$(n + 2)^d - n^d \ = \ (1 + 2)^d - 1^d \ = \ 3^d - 1.$$  (5)

The right-hand member of (5) gives us a formula for the total number of sub-elements of a **d**-duplex. If we include the **d**-duplex together with its sub-elements, we have a total of $3^d$ elements. For example, in the case of the **4**-duplex (tesseract), the right-hand member of (5) is $3^4 - 1 = 80$, which is the total number of sub-elements of the tesseract. At the same time, for d = 4 and n = 1, the right-hand member of (1) also evaluates to 80,

$$4 \cdot 2 + 6 \cdot 2^2 + 4 \cdot 2^3 + 1 \cdot 2^4 = 8 + 24 + 32 + 16$$  (6)
$$= 80.$$

The values of the 4 terms of the right-hand member of (6) gives us the number of each of the 4 types of sub-elements of a tesseract. That is, the number and type of sub-elements of a tesseract are

$$N(4, 3) = \ 8 \ \text{cells (cubes)}$$
$$N(4, 2) = 24 \ \text{faces (squares)}$$
$$N(4, 1) = 32 \ \text{edges (line-segments)}$$
$$N(4, 0) = \underline{16} \ \text{vertices (points)}$$
$$\text{Total} \qquad 80$$

In general, the sum of the row elements of the Duplex Triangle is given by

$$\sum N(d, k) = 3^d - 1, \quad \text{for } k = 0 \text{ to } d - 1,$$  (7a)

and,

$$\sum N(d, k) = 3^d, \quad \text{for } k = 0 \text{ to } d.$$  (7b)

    **Section 2.3**

1.    Use the formula $3^d - 1$ to find the number of sub-elements of the **5**-duplex.

2.        Find the number of sub-elements of the **6**-duplex.

---

**Alternating sums of the numbers of sub-elements and Euler's Formula**
Furthermore, if we let n = 1 in the right-hand member of (1) and alternately add and subtract the numbers of each type of sub-element, we get 0 if d is even and we get 2 if d is odd.  For example, for d = 3, we have,

$$3 \cdot 2 \cdot n^2 - 3 \cdot 2^2 \cdot n^1 + 1 \cdot 2^3 \cdot n^0 = 3 \cdot 2 - 3 \cdot 4 + 1 \cdot 8$$
$$= 6 - 12 + 8$$
$$= 2$$

This, of course, is Euler's result for a cube (and all convex polyhedrons),  F – E + V = 2 (**Faces – Edges + Vertices = 2**), or equivalently,

$$V - E + F = 2. \qquad \text{(Euler's Formula)}$$

Alternating  the signs  in the  right-hand  member of (1)  for d = 4  (the tesseract) and n = 1, we have

$$4 \cdot 2 - 6 \cdot 2^2 + 4 \cdot 2^3 - 1 \cdot 2^4 = 8 - 24 + 32 - 16 = 0.$$

These **Eulerian formulas** hold for all **d**-duplexes, d ≥ 1.  The proof is as follows:

To obtain an expression for the right-hand member of (1) with alternating signs, we introduce a  factor of (–1) in the left-hand  member of (1) by letting n = – 1 and then expanding the binomial, giving us (8).

$$(-1 + 2)^d - (-1)^d = C(d, 1) \cdot 2 \cdot (-1)^{(d-1)} + C(d, 2) \cdot 2^2 \cdot (-1)^{(d-2)}$$
$$+ C(d, 3) \cdot 2^3 \cdot (-1)^{(d-3)}$$
$$+ \ldots + C(d, d-1) \cdot 2^{(d-1)} \cdot (-1)^{(d-(d-1))}$$
$$+ C(d, d) \cdot 2^d \cdot (-1)^{(d-d)} \qquad (8)$$

Now, if d is even, we have (by evaluating the left member of (8))

$$0 = - C(d, 1) \cdot 2 + C(d, 2) \cdot 2^2 - C(d, 3) \cdot 2^3$$
$$+ \ldots + (-1)^{(1)} \cdot C(d, d-1) \cdot 2^{(d-1)}$$
$$+ (-1)^{(0)} \cdot C(d, d) \cdot 2^d , \qquad (9)$$

and if d is odd, we have

$$2 = C(d, 1) \cdot 2 - C(d, 2) \cdot 2^2 + C(d, 3) \cdot 2^3$$
$$+ \ldots + (-1)^{(1)} \cdot C(d, d-1) \cdot 2^{(d-1)}$$
$$+ (-1)^{(0)} \cdot C(d, d) \cdot 2^d . \qquad \blacksquare \qquad (10)$$

*Note: If we multiply both sides of (9) by –1, we will still get 0 on the left-hand side and alternating signs on the right-hand side, but starting with a positive first term.*

The next table shows the interesting connection between the rows of Pascal's Triangle and the powers of 2 that combine to determine the number of sub-elements of each **d**-duplex for d = 1 to 5. The entries in the triangular array shown in the table are equal to N(d, k), where d is the row number and k is the column number, beginning with k = 0. The alternating sum of the entries in each row equals 2 if d is odd, or 0 if d is even.

| d-duplex | Number of Sub-elements | | | | | Total $3^d - 1$ | Alternating Sum |
|---|---|---|---|---|---|---|---|
| | 0-D | 1-D | 2-D | 3-D | 4-D | | |
| 1-Duplex (Line Segment) | 2•1 | | | | | 2 | 2 |
| 2-Duplex (Square) | 4•1 | 2•2 | | | | 8 | 0 |
| 3-Duplex (Cube) | 8•1 | 4•3 | 2• 3 | | | 26 | 2 |
| 4-Duplex (Tesseract) | 16•1 | 8•4 | 4• 6 | 2• 4 | | 80 | 0 |
| 5-Duplex | 32•1 | 16•5 | 8•10 | 4•10 | 2•5 | 242 | 2 |

If we include the original **d**-duplex as another element, along with its sub-elements, then starting with the number of **0**-duplexes (vertices) then subtracting the number of **1**-duplexes (edges), then adding the number of **2**-duplexes (faces), and continuing to alternate the adding and subtracting of the number of **k**-duplexes up to the original single **d**-duplex, the result will always equal 1. In general,

$$\mathbf{N(d, 0) - N(d, 1) + N(d, 2) \ - \ ... \ + (-1)^{d-1} \cdot N(d, d-1)] + (-1)^d \cdot N(d, d) \ = \ 1} \qquad (11)$$

or, equivalently,

$$\sum (-1)^k \cdot N(d, k) \ = \ 1, \ \text{ for } k = 0 \text{ to } d. \qquad \text{Note: } N(d, d) = 1.$$

We will say that **Euler's formula for duplexes** is the equivalent of (11).

In the case of the **3**-duplex (cube), we have $8 - 12 + 6 - 1 = 1$ and in the case of the **4**-duplex (tesseract), we have $16 - 32 + 24 - 8 + 1 = 1$.

**Proof of (11):**

Because N(d, k) is equal to the number of **k**-duplexes (sub-elements) of the **d**-duplex (by definition given in (4)), the first d terms of the left member of (11) are the same as the d terms on the right-hand side of (8), but displayed in reverse order. N(d, d), in the last term of the left member of (11) (representing the **d**-duplex itself as an element) is equal to 1.

So, if d is even, we have

$$\mathbf{[N(d, 0) - N(d, 1) + N(d, 2) \ - \ ... \ + (-1)^{d-1} \cdot N(d, d-1)] + (-1)^d \cdot N(d, d)} \ = 0 + 1 = 1$$
because of (9) and N(d, d) = 1.

And if d is odd, we have

$$\mathbf{[N(d, 0) - N(d, 1) + N(d, 2) \ - \ ... \ + (-1)^{d-1} \cdot N(d, d-1)] + (-1)^d \cdot N(d, d)} \ = 2 - 1 = 1$$
because of (10) and N(d, d) = 1. ∎

---

3.    Use the formulas, equations (7) and (11), to find the sum of the elements of the **5**-duplex and the alternating sum of those numbers.

The table below extends the previous table in this section to include the **d**-duplex as the final entry of 1 in each row (this is the same table presented in Sec. 1.4). The numbers (bold) in the table are given by the formula for N(d, k), equation (4), where d is the dimension of the duplex and k is the dimension of a specified sub-element, and these are the numbers in the rows of the Duplex Triangle.

| d-duplex | Number of Elements of each Type | | | | | | Total $3^d$ | Alternating Sum |
|---|---|---|---|---|---|---|---|---|
| | 0-D | 1-D | 2-D | 3-D | 4-D | 5-D | | |
| **0**- Duplex (Point) | **1** | | | | | | 1 | 1 |
| **1**- Duplex (Line-Segment) | **2** | **1** | | | | | 3 | 1 |
| **2**- Duplex (Square) | **4** | **4** | **1** | | | | 9 | 1 |
| **3**- Duplex (Cube) | **8** | **12** | **6** | **1** | | | 27 | 1 |
| **4**- Duplex (Hypercube) | **16** | **32** | **24** | **8** | **1** | | 81 | 1 |
| **5**- Duplex | **32** | **80** | **80** | **40** | **10** | **1** | 243 | 1 |

**Table of Values of N(d, k)**

Note: A formula similar to (11) for other convex polytopes in d-space, including all three of the strands of regular convex polytopes being treated in this text, also holds, yielding a value of 1 for the alternating sum, see [1], Chap. IX.

---

4.    Refer to the table of simplexes near the end of Section 1.2 and find the alternating sum of those numbers for the **3**-simplex, **4**-simplex, and the **5**-simplex.

---

The formula for N(d, k), equation (4), also gives us the entry in the Duplex Triangle in row d and column k. The entry N(4, 2) = C(4, 2) • $2^{4-2}$ = 6 • 4 = 24 is circled in the triangular array below; it is in row 4 and column 2.

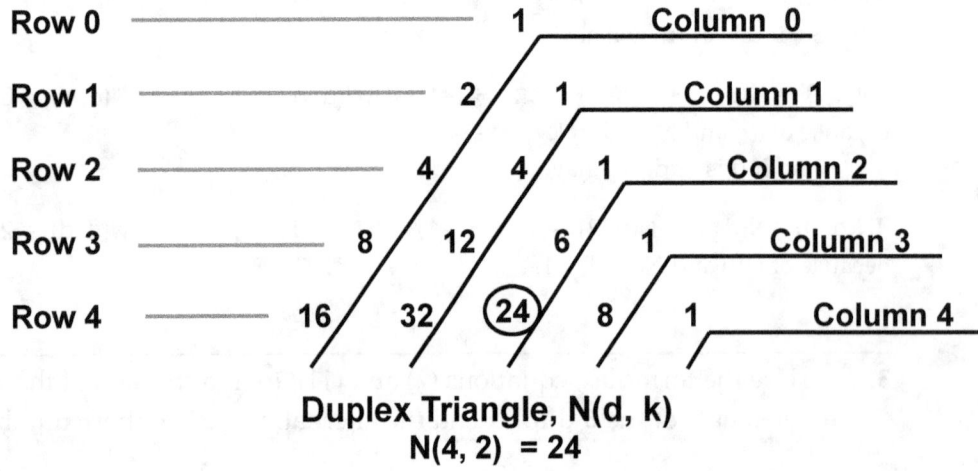

**Duplex Triangle, N(d, k)**
**N(4, 2) = 24**

5.      Use the formula, equation (4), to find the entry in row 3 and column 1 of the Duplex Triangle.

## 2.4   The Duplex Triangle Rule

We also have a rule for the Duplex Triangle, that is similar to Pascal's Rule for Pascal's Triangle. Namely,

$$\boxed{N(d-1, k-1) \; + 2 \cdot N(d-1, k) \; = \; N(d, k).}$$      (12)

In words, equation (12) states that the number of sub-elements of the type **(k − 1)**-duplex contained in a **(d − 1)**-duplex, plus two times the number of sub-elements of the type **k**-duplex is equal to the number of sub-elements of the type **k**-duplex contained in a **d**-duplex.

**Proof of (12):**

     Equation (4) defines the expression $N(d, k) = C(d, d-k) \cdot 2^{d-k} = C(d, k) \cdot 2^{d-k}$.

     Substituting $C(d, k) \cdot 2^{d-k}$ for $N(d, k)$ into the left member of (12), we have

$$C(d-1, k-1) \cdot 2^{(d-1)-(k-1)} + 2 \cdot C(d-1, k) \cdot 2^{(d-1)-k}$$
$$= \; C(d-1, k-1) \cdot 2^{d-k} + 2 \cdot C(d-1, k) \cdot 2^{(d-k)-1}$$
$$= \; C(d-1, k-1) \cdot 2^{d-k} + C(d-1, k) \cdot 2^{(d-k)}$$
$$= \; 2^{d-k} \cdot [C(d-1, k-1) + C(d-1, k)]$$

and by Pascal's Rule we have

$$= \; 2^{d-k} \cdot [C(d, k)]$$
$$= \; N(d, k),$$

which is the right-hand member of (12).      ∎

     In Chap. 1 (the discussion of geometric duplication) we stated a specific instance of the Duplex Triangle Rule. There, we stated that the number of edges of a **d**-duplex is two times the number of edges of the **(d − 1)**-duplex plus the number of vertices of the **(d − 1)**-duplex.

     **An Example.** Let $d = 5$ and $k = 3$. Then the left-hand side of (12) becomes

$$N(4, 2) \; + \; 2 \cdot N(4, 3) = C(4, 2) \cdot 2^2 + 2 \cdot (C(4, 3) \cdot 2^1)$$
$$= 6 \cdot 4 + 2 \cdot (4 \cdot 2)$$
$$= 24 + 2 \cdot 8 = 40$$

and the right-hand side of (12) also equals 40,

$$N(5, 3) = C(5, 3) \cdot 2^2 = 10 \cdot 4 = 40.$$

The positions of these entries, that is 24, 8 and 40 in the previous example, are highlighted in the array shown below.

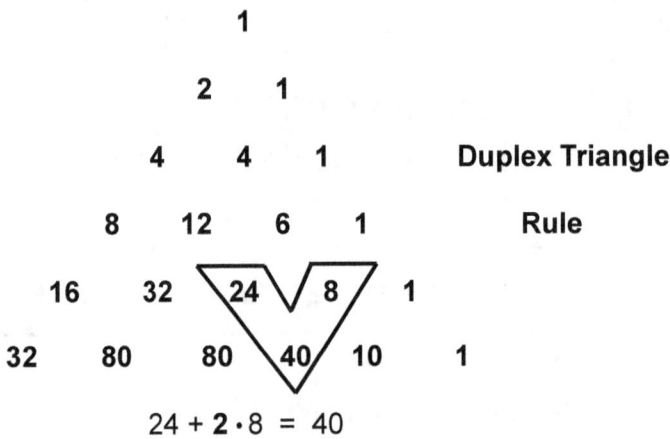

$$24 + 2 \cdot 8 = 40$$

The rule, equation (12), like Pascal's Rule, is a recursive rule. That is, if we know the entries of row $d - 1$, then we can find the entries for row $d$ from row $d - 1$.

### Section 2.4

1.    Use the recursive rule for the Duplex Triangle to find the entries of $6^{th}$ row of the Triangle, given the entries of the $5^{th}$ row are as shown in the triangular array above.
2.    Find the entries of the $7^{th}$ row of the Duplex Triangle.

## 2.5  Extending The Duplex Triangle Rule

In this section we extend the Duplex Triangle rule, in a manner similar the the extension of the Pascal Triangle rule. To characterize these extensions we define a new operation called a **"string dot product"**.   Recall, a string product result is a  string (see Sec. 2.1).  For example,

$$(1, 3, 3, 1)(5, 10, 10, 5) = (5, 30, 30, 5)$$

A **string dot product** result, is a sum of a sequence of the products of pairs of numbers in corresponding positions of the operand strings.  To find the dot product of two strings, for example the string corresponding to the $3^{rd}$ row of Pascal's Triangle, $(1, 3, 3, 1)$, and the string corresponding to the four middle elements of the $5^{th}$ row of Pascal's Triangle, $(5, 10, 10, 5)$, we write

$$
\begin{aligned}
(1, 3, 3, 1) \bullet (5, 10, 10, 5) &= 1 \bullet 5 + 3 \bullet 10 + 3 \bullet 10 + 1 \bullet 5 \qquad \Longleftarrow \quad \text{a sum of products} \\
&= \phantom{0}5 \; + \; 30 \; + \; 30 \; + 5 \qquad\qquad \Longleftarrow \quad \text{intermediate result} \\
&= \phantom{0}70 \qquad\qquad\qquad\qquad\qquad\qquad \Longleftarrow \quad \text{string dot product}
\end{aligned}
$$

Furthermore, the result, 70, is the middle element in the $8^{th}$ row of Pascal's Triangle. The row number, 8, is the sum of the row numbers of the strings forming the dot product, $3 + 5 = 8$. See the figure below.

```
                1
              1   1
            1   2   1
          ( 1   3   3   1 )
          1   4   6   4   1
        1 ( 5   10  10  5 ) 1
        1   6   15  20  15  6   1
      1   7   21  35  35  21  7   1
    1   8   28  56 (70) 56  28  8   1      C(8, 4) = 70
  1   9   36  84  126 126 84  36  9   1
```

In general, if we form the string dot product of elements from the n-th and m-th rows, we get the element in the (n + m)-th row and the c-th column (which is generally the column position of the right-most entry of the second string).

> (Note: The sum of products is very pervasive and useful in many aspects of mathematics. In the study of vector algebra, the string dot product operation is simply called the "dot product" and in that context our strings would be regarded as vectors.)

There are similar results when finding string dot products using the rows of the Duplex Triangle. However, in the case of the Duplex Triangle, we must reverse the order of the entries of the first string operand. For example, consider the string dot product using the second row of the Duplex Triangle and the middle elements from the $4^{th}$ row.

```
                    1
                  2   1
  Reverse this row → ( 4   4   1 )
                8   12  6   1
            16 ( 32  24  8 ) 1
          32  80  80  40  10  1
        64  192 240 (160) 60  12  1
```

$$(1, 4, 4) \bullet (32, 24, 8) = 1 \bullet 32 + 4 \bullet 24 + 4 \bullet 8$$
$$= 32 + 96 + 32$$
$$= 160$$

$N(6, 3) = 160$

The result, 160, is the element in the $6^{th}$ row ($2 + 4 = 6$) and the $3^{rd}$ column (the element 8 in the second string is the $3^{rd}$ element in the $4^{th}$ row.

Pascal's Rule can be characterized as the string representing row 1, (1, 1), multiplied by the string of two consecutive entries in another row of Pascal's Triangle, such as, (4, 6), found in the 4$^{th}$ row of Pascal's Triangle, to generate an element in the row 5.

$$(1, 1) \bullet (4, 6) = 4 + 6 = 10$$

which is the element $C(5, 2) = 10$ in the 5$^{th}$ row of Pascal's Triangle $(5 = 1 + 4)$.

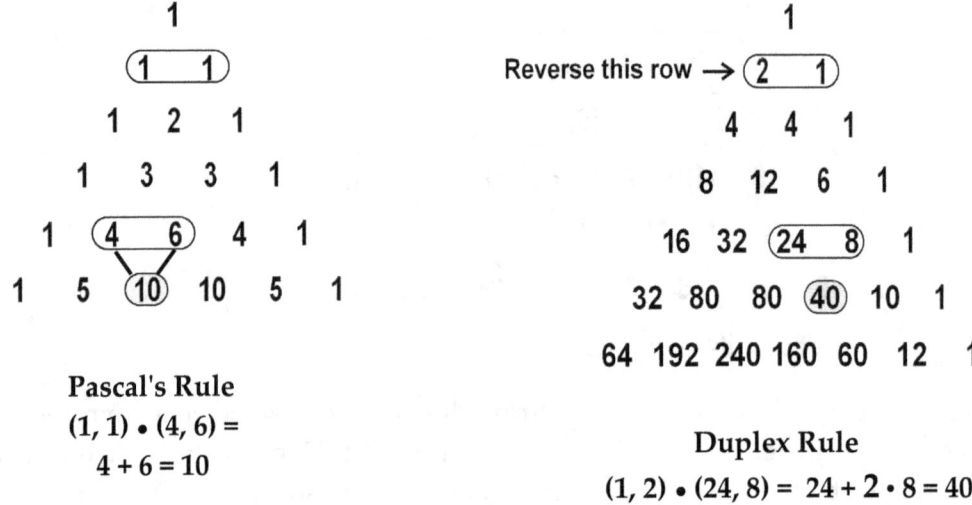

**Pascal's Rule**
**(1, 1) • (4, 6) =**
**4 + 6 = 10**

**Duplex Rule**
**(1, 2) • (24, 8) = 24 + 2 • 8 = 40**

The Duplex Rule can be characterized as the reverse of the string (2, 1), that is, row 1, multiplied by the string of two consecutive entries in another row of the Duplex Triangle, such as, (24, 8), found in the 4$^{th}$ row of the Duplex Triangle, to generate an element in row 5.

$$(1, 2) \bullet (24, 8) = 24 + 2 \bullet 8 = 24 + 16 = 40 = N(5, 3)$$

In order to discuss the extensions of these rules and the string dot products of different rows of the Triangles, we will assume that each row consists of its normal elements and as many additional zeros as necessary to ensure that both strings have an equal number of elements. Thus, to form the string dot product of the 3$^{rd}$ and 6$^{th}$ rows of Pascal's Triangle, we write

$$
\begin{aligned}
(1, 3, 3, 1, 0, 0, 0) \bullet (1, 6, 15, 20, 15, 6, 1) &= 1 \bullet 1 + 3 \bullet 6 + 3 \bullet 15 + 1 \bullet 20 + 0 \bullet 15 + 0 \bullet 6 + 0 \bullet 1 \\
&= 1 + 18 + 45 + 20 + 0 + 0 + 0 \\
&= 84
\end{aligned}
$$

The result of this string dot product is 84, which is the element in the 9$^{th}$ row and 6$^{th}$ column of Pascal's Triangle, $C(9, 6) = 84$. We get the same result multiplying

$$(1, 3, 3, 1) \bullet (1, 6, 15, 20) = 84,$$

but, this time the result is $C(9, 3)$, which also equals 84. That is, $C(9, 3) = C(9, 6)$, because of the symmetry in Pascal's Triangle. In general,

$$C(n + m, n) = C(n + m, m).$$

Incidentally, the sum of the squares of the elements in a row of Pascal's Triangle is the string dot product of the row with itself.

$$
\begin{aligned}
(1, 3, 3, 1) \bullet (1, 3, 3, 1) &= 1 \bullet 1 + 3 \bullet 3 + 3 \bullet 3 + 1 \bullet 1 \\
&= 1^2 + 3^2 + 3^2 + 1^2 \\
&= 20 \\
&= C(6, 3)
\end{aligned}
$$

For the string dot products of different rows of the Duplex Triangle, we will also assume that each row consists of its normal elements and as many additional zeros as necessary to ensure that the strings have an equal number of elements. However, in the case of the Duplex Triangle, we will usually reverse the string of the smaller numbered row, when forming string dot products. Thus, to form the string dot product of the 2nd and 4th rows of the Duplex Triangle, we write row 2 with appended zeroes and then reverse the row as follows:

$$
\begin{aligned}
(0, 0, 1, 4, 4) \bullet (16, 32, 24, 8, 1) &= 0 \bullet 16 + 0 \bullet 32 + 1 \bullet 24 + 4 \bullet 8 + 4 \bullet 1 \\
&= 0 \quad + \quad 0 \quad + 24 + 32 \quad + 4 \\
&= 60
\end{aligned}
$$

The result of this string dot product is 60, which is the element in the 6th row (2 + 4 = 6) and 4th column of the Duplex Triangle. N(6, 4) = 60.

In general,

*if we form the string dot product of the n-th and m-th rows of the Duplex Triangle, as described above, we get the element in the (n + m)-th row and the m-th column of the Duplex Triangle.*

```
                      1

                   2   1

Reverse this row →  ⟨4   4   1   0   0⟩

                  8  12   6   1

                 ⟨16  32  24   8   1⟩

              32  80  80  40  10   1

           64  192 240 160 ⟨60⟩ 12   1        N(6, 4) = 60
```

**(0, 0, 1, 4, 4) • (16, 32, 24, 8, 1) = 0 + 0 + 24 + 32 + 4 = 60**

**Section 2.5**

1. Compute this string dot product:    $(1, 2, 1) \bullet (4, 6, 4) = ?$
   In what row and column can the result be found in Pascal's Triangle?

2. Compute this string dot product:    $(1, 4, 6, 4, 1) \bullet (1, 4, 6, 4, 1) = ?$
   In what row and column can the result be found in Pascal's Triangle?

3. What is the sum of the squares of the elements in the 2nd row of Pascal's Triangle?

4. What is the sum of the squares of the elements in the 5th row of Pascal's Triangle?  In what row and column can the result be found in Pascal's Triangle?

5. Compute the string dot product of the 2nd row of the Duplex Triangle with itself (reverse the order of the elements in the first string).  In what row and column can the result be found in the Duplex Triangle?

6. Compute the string dot product of the two strings shown in Pascal's Triangle below.

   In what row and
   column can the
   result be found?

   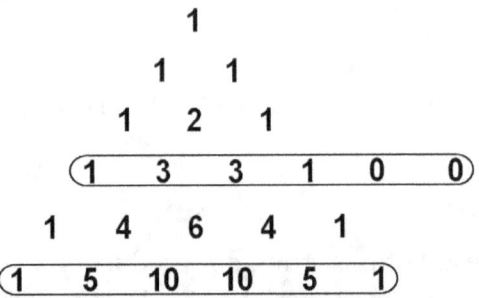

7. Compute the string dot product of the two strings shown in the Duplex Triangle below.  In what row and column can the result be found in the Duplex Triangle?

```
                1
             2    1
Reverse this row → (4    4    1    0)
             (8   12    6    1)
          16   32   24    8    1
        32   80   80   40   10   1
      64  192  240  160   60   12   1
```

8. Given a cube.  Let **V** = the number of vertices, **E** = the number of edges, **F** = the number of faces, and **C** = the number of cubes.    ⟶

    (a)  Compute the string dot product **(C, F, E, V) • (V, E, F, C)**, first in symbols, then in numbers.

    (b)  Show that **2(V + EF) = N(6,3)**, that is, show that twice the sum of the the number of vertices and the product of the number of edges and faces of a cube, is equal to the number of cubic sub-elements of a **6**-duplex.

## Summary:

**Duplex** is the general term of the sequence line-segment, square, cube, hypercube, . . . . Each subsequent term in the sequence is of a dimension one greater than its predecessor.

A **d-duplex** is a d-dimensional regular convex polytope consisting of sub-elements of the type points (vertices); line-segments (edges); squares (faces); cubes; hypercubes; and higher dimensional sub-elements (cells) up to **(d − 1)**-duplexes.

The number of sub-elements of a **d**-duplex of a given type, the type **k**-duplex, $0 \leq k \leq d$, is represented by the expression N(d, k). These numbers are the entries in row d, column k of the **Duplex Triangle.**

**N(d, k) = C(d, d − k) • $2^{d-k}$ = C(d, k) • $2^{d-k}$,** the entries in row d, column k of the **Duplex Triangle,** where $0 \leq k \leq d$ and C(d, k) is the binomial coefficient, C(d, k) = d!/(k! • (d − k)!).

**N(d, d) = 1,** N(d, d) represents the final entry in each row of the Duplex Triangle and corresponds to the **d**-duplex itself, which is always 1.

**N(d, 0) = $2^d$**
N(d, 0) represents the number of verticies of the **d**-duplex, which are the entries in column 0 of the Duplex Triangle. The number of verticies in the **d**-duplex is double the number of vertices in the **(d − 1 )**-duplex, due to its formation from the **(d − 1 )**-duplex by the process of geometric duplication.

**N(d, d − 1) = 2d**
After the formation of the **d**-duplex from the **(d − 1 )**-duplex, by the process of geometric duplication, the number of **(d − 1 )**-duplexes in the **d**-duplex is 2 times the dimension of the newly formed **d**-duplex. This can be observed in the Duplex Triangle by examining the entries just before the final 1 in each row.

**$\sum$N(d, k) = $3^d$ − 1,** for k = 0 to d − 1, or, **$\sum$N(d, k) = $3^d$,** for k = 0 to d
The sum of the numbers of sub-elements of a **d**-duplex is equal to $3^d$ − 1, or, $3^d$, if the **d**-duplex itself is included in the total. The row total for each row of the Duplex Triangle is $3^d$. (Note: The row total for each row of Pascal's Triangle is $2^d$.)

**Euler's Formula:  V − E + F = 2**         (the alternating sum of the numbers of sub-elements of a regular convex polyhedron, but not including the polyhedron itself)

$\sum (-1)^k \cdot N(d, k) = 1$,  for k = 0 to d.

The alternating sum of the numbers of sub-elements of a  **d**-duplex, including the **d**-duplex itself, is equal to 1.   The alternating sum of the elements in each row of the Duplex Triangle is 1.  (Note:  The alternating sum of the elements in each row (row 1 and greater) of Pascal's Triangle is 0, but the alternating sum of the elements in each row of the Simplex Triangle is 1, that is, the alternating sum of the numbers of sub-elements of a  **d**-simplex, including the **d**-simplex itself, is equal to 1. )

**N(d − 1, k − 1)  + 2 · N(d − 1, k)  =  N(d, k)**

The Duplex Triangle Rule.  Similar to Pascal's Rule for Pascal's Triangle, except here the second element is multiplied by 2.

Pascal's Rule:  **C(d − 1, k − 1)  +  C(d − 1, k)  =  C(d, k).**

|                              | 1   |     |     |     |     |     |     |
|------------------------------|-----|-----|-----|-----|-----|-----|-----|
|                              | 2   | 1   |     |     |     |     |     |
| **Duplex Triangle**          | 4   | 4   | 1   |     |     |     |     |
| **(rows 0 through 6)**       | 8   | 12  | 6   | 1   |     |     |     |
|                              | 16  | 32  | 24  | 8   | 1   |     |     |
|                              | 32  | 80  | 80  | 40  | 10  | 1   |     |
|                              | 64  | 192 | 240 | 160 | 60  | 12  | 1   |

$$2^7 \cdot 1 \quad 2^6 \cdot 7 \quad 2^5 \cdot 21 \quad 2^4 \cdot 35 \quad 2^3 \cdot 35 \quad 2^2 \cdot 21 \quad 2^1 \cdot 7 \quad 2^0 \cdot 1$$

**Row 7 of the Duplex Triangle (in factored form).**
**The string product of the powers of 2 from the 7th down to the 0th power**
**with the string representing the 7th row of Pascal's Triangle.**

# 3

# Number Patterns Within the Duplex Triangle

## 3.1  Some Number Patterns Related to Duplex Triangle

| d | N(d, k) | | | | | | | Total | Alt. Sum |
|---|---|---|---|---|---|---|---|---|---|
| 0 | 1 | | | | | | | $1 = 3^0$ | 1 |
| 1 | 2 | 1 | | | | | | $3 = 3^1$ | 1 |
| 2 | 4 | 4 | 1 | | | | | $9 = 3^2$ | 1 |
| 3 | 8 | 12 | 6 | 1 | | | | $27 = 3^3$ | 1 |
| 4 | 16 | 32 | 24 | 8 | 1 | | | $81 = 3^4$ | 1 |
| 5 | 32 | 80 | 80 | 40 | 10 | 1 | | $243 = 3^5$ | 1 |
| 6 | 64 | 192 | 240 | 160 | 60 | 12 | 1 | $729 = 3^6$ | 1 |

Column 0 ⟶

**The Duplex Triangle, N(d, k)**

In Chapter 1, we established that the sum of the entries in row d was equal to a power of 3, $3^d$. In Chapter 2, we established that the alternating sum of the entries in row d was equal to 1. We also developed the formula for the element in the d-th row and k-th column of the Duplex Triangle, namely,

$$N(d, k) = C(d, d - k) \cdot 2^{d-k} = C(d, k) \cdot 2^{d-k},$$

where $C(d, k)$ is the binomial coefficient, the element in row d, column k of Pascal's Triangle.

The Duplex Triangle Rule is a recursive rule for the element in the d-th row and k-th column of the Triangle. It is similar to Pascal's Rule for Pascal's Triangle, except here the second element in the right-hand member is multiplied by 2.

$$N(d, k) = N(d - 1, k - 1) + 2 \cdot N(d - 1, k)$$

In the Duplex Triangle shown above, the Duplex Triangle Rule states that a given entry in row d is equal to the sum of two times the entry directly above the given entry, row $d - 1$, and the entry that precedes it.

In the Duplex Triangle shown above, observing column 0, we see the pattern

$$N(d, 0) = 2^d$$

representing the number of verticies of the **d**-duplex.  The number of verticies in the **d**-duplex is double the number of vertices in the **(d – 1 )**-duplex, due to its formation from the **(d – 1 )**-duplex by the process of geometric duplication.

After the formation of the **d**-duplex from the **(d – 1 )**-duplex, by the process of geometric duplication, the number of boundary elements, **(d – 1 )**-duplexes, in the **d**-duplex is 2 times the dimension of the newly formed **d**-duplex.  That is,

**N(d, d – 1) = 2d**

This can be observed in the Duplex Triangle by examining the entries just before the final 1 in each row.  See Fig. 1.

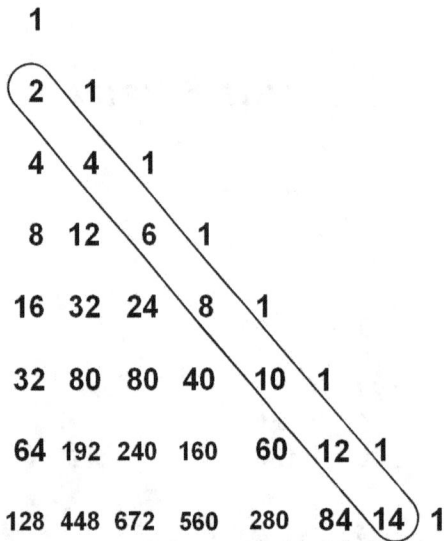

Figure 1.   N(d, d – 1) = 2d
**The encircled elements belong to a cross-column of the Duplex Triangle (see Fig. 2).**

Another property, referring to the Duplex Triangle above, relates to even numbered rows.  The element in the middle of an even numbered row is 4 times the element in preceding row and just above the given middle element.   For example, in row 6 the middle element is 160.  The element in the preceding row and just above 160 is 40 and 4 · 40 = 160.  In general,

**N(2x, x)  =  4 · N(2x – 1, x),   for x = 1, 2, 3, . . .**

For the special case x = 3, in the formula above, it states that the number of sub-elements of type **3**-duplex, that is cubes, contained in a **6**-duplex is 4 times the number of cubes contained in a **5**-duplex.

This next property relates to the odd numbered rows.  Consider the two elements nearest the middle of the row.  One-half of the element on the left equals the element on the right.   For example, in row 7, the two elements nearest the middle of the row are N(7, 3) =  560 and N(7, 4) = 280.  (½) · N(7, 3) = N(7, 4).  In general,

**(½) · N(2x + 1, x)  =  N(2x + 1, x + 1)**

In the exercises, Section 3.1, we will discover some other interesting properties of the numbers in the Duplex Triangle.

          **Section 3.1**

1.      Perform the indicated computations in the rows of the Duplex Triangle shown below and complete the following equations:

$$1 \qquad = \mathbf{1}$$

$$2 + 1 \qquad = \mathbf{3}$$

$$4 + 4 - 1 \qquad = \underline{\qquad}$$

$$8 + 12 - 6 + 1 \qquad = \underline{\qquad}$$

$$16 + 32 - 24 + 8 - 1 = \underline{\qquad}$$

$$32 + 80 - 80 + 40 - 10 + 1 = \underline{\qquad}$$

$$64 + 192 - 240 + 160 - 60 + 12 - 1 = \underline{\qquad}$$

Describe the sequence of numbers generated. (Hint: If you do not recognize the numbers, try adding 1 to each number to help you identify them.)

2.      On each row of the Duplex Triangle, shown below, perform the indicated computations, not including the elements enclosed in the box. Describe the sequence of resulting numbers.

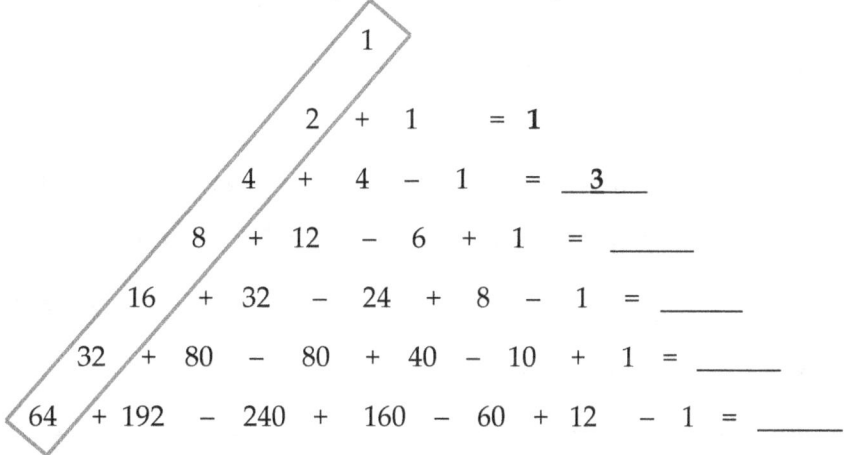

3.      Verify the formula  $\mathbf{N(d, k) = N(d - 1, k - 1) + 2 \cdot N(d - 1, k)}$  for d = 4 and k = 2.

4.      Verify the formula  $\mathbf{N(2x, x) = 4 \cdot N(2x - 1, x)}$, **for x = 2.** That is, show that

$$\mathbf{N(4, 2) = 4 \cdot N(3, 2)}$$

5.      Develop row 8 of the Duplex Triangle. Show that

$$\mathbf{N(8, 4) = 4 \cdot N(7, 4)}$$

6.    Verify the formula   $(\frac{1}{2}) \cdot N(2x + 1, x) = N(2x + 1, x + 1)$,   for x = 2.
That is, show that

$$(\frac{1}{2}) \cdot N(5, 2) = N(5, 3)$$

7.    Show that (a)   $(\frac{1}{2}) \cdot N(3, 1) = N(3, 2)$   and  (b) $(\frac{1}{2}) \cdot N(7, 3) = N(7, 4)$.

(c)  Choose any number, x, greater than 3, and verify the formula

$$(\frac{1}{2}) \cdot N(2x + 1, x) = N(2x + 1, x + 1).$$

8.    Choose any row of the Duplex Triangle, for example the $5^{th}$ row.
Separate the elements into two rows as shown.

|  |  |  |  |  |  |  | 80 | 40 | 1 |
|---|---|---|---|---|---|---|---|---|---|
| 32 | 80 | 80 | 40 | 10 | 1 | ⟹ | 32 | 80 | 10 |

Form a fraction with the sum of the numbers in the top row as the
numerator and the sum of the numbers in the bottom row as the
denominator.  Add 1 to the numerator.

$$\frac{80 + 40 + 1 + \mathbf{1}}{32 + 80 + 10}$$

The value of this fraction is 1.
Follow the process described above for rows 1 through 6 of the Duplex
Triangle.  What are the values of these fractions?

9.    Verify the formula   $x \cdot N(x + 1, x) = N(x + 1, x - 1)$,  for x = 1, 2, 3 and 4.

10.    Show that  $x \cdot N(x + 1, x) = 2x^2 + 2x$.

11.    (a)  Verify the formula   $x \cdot N(x - 1, 0) = N(x, 1)$,  for x = 1, 2, 3 and 4.

(b)  Verify the formula   $x \cdot 2^{x-1} = N(x, 1)$,  for x = 1, 2, 3 and 4.

12.    (a) Verify the formula  $2^{x-1} \cdot N(x + 1, x) = N(x + 1, 1)$,  for x = 1, 2, 3 and 4.

(b)  Verify the formula  $2^x \cdot (x + 1) = N(x + 1, 1)$,  for x = 1, 2, 3 and 4.

13.    Verify the formula  $2^{x-2} \cdot N(x + 2, x) = N(x + 2, 2)$,  for x = 1, 2, 3, 4 and 5.

14.    Verify the formula  $2^{x-3} \cdot N(x + 3, x) = N(x + 3, 3)$,  for x = 1, 2, 3, 4 and 5.

15.    Line up the formulas from Exercises 12(a), 13, and 14.  Predict the next
formula for that sequence.

___

Like Pascal's Triangle, there are many number patterns in the Duplex Triangle.  We have explored a few here and we are confident that there are many more.  In the next Section we will explore additional patterns involving the columns and cross-columns of the Duplex Triangle.

## 3.2 Number Patterns Related to the Cross-columns of the Duplex Triangle

Since the Duplex Triangle does not have the symmetry found in Pascal's Triangle, we will examine columns in both directions across the Triangle. The second set of columns will be referred to as "**cross-columns**".

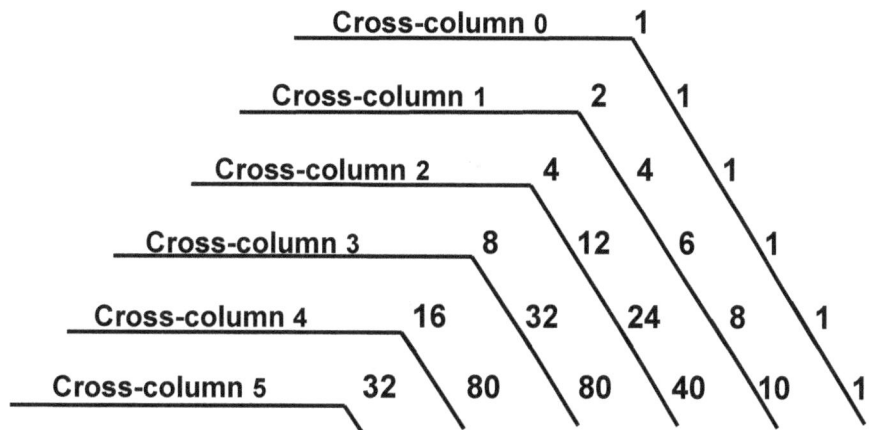

**Figure 2. Illustrating the cross-columns of the Duplex Triangle**

Consider the sums of successive cross-column elements. For example, the cross-column 2 begins as follows:

4

12

24

40

60

Now form a sequence of partial sums in the following way:

$$4 = 4 = (\tfrac{1}{2}) \cdot 8$$
$$4 + 12 = 16 = (\tfrac{1}{2}) \cdot 32$$
$$4 + 12 + 24 = 40 = (\tfrac{1}{2}) \cdot 80$$
$$4 + 12 + 24 + 40 = 80 = (\tfrac{1}{2}) \cdot 160$$
$$4 + 12 + 24 + 40 + 60 = 140 = (\tfrac{1}{2}) \cdot 280$$

Observe the pattern of partial sums: 4, 16, 40, 80, 140. If we double these numbers, we get the beginning of the elements in cross-column 3, that is, 8, 32, 80, 160, 280. We can write this sum as follows:

$$N(2, 0) + N(3, 1) + N(4, 2) + N(5, 3) + N(6, 4) = (\tfrac{1}{2}) \cdot N(7, 4)$$

In general, the sum of the first n + 1 elements of cross-column 2 is equal to one-half the n-th element in cross-column 3. That is,

**N(2, 0) + N(3, 1) + N(4, 2) + N(5, 3) + . . . + N(n + 2, n) = (½) • N(n + 3, n)**       (1)

This is illustrated in the Duplex Triangle below for n = 4.

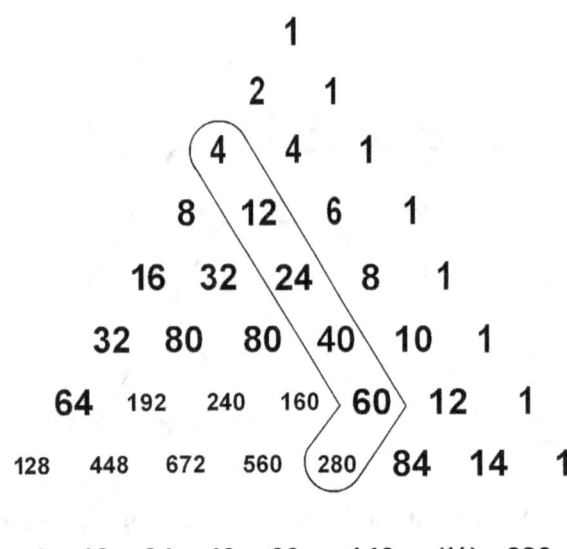

**4 + 12 + 24 + 40 + 60 = 140 = (½) • 280**

**Figure 3.  A property of the cross-columns of the Duplex Triangle,**
**that is similar to the "hockey-stick" property of Pascal's Triangle (see Fig. 4).**

       **Section 3.2**

1.    (a)  What is the sum of the first 3 elements of cross-column 2?
      (b)  What is the sum of the first 4 elements of cross-column 2?

2.    Let n = 5 in equation (1) above.   Find the sum of the first 6 elements
      of cross-column 2. If you double the sum, in what row and column
      do you find this value?

3.    Encircle the first 5 elements of cross-column 1 and find their sum.
      If you double the sum, in what row and column do you find this value?

4.    Encircle the first 3 elements of cross-column 4 and find their sum.
      If you double the sum, in what row and column do you find this value?

5.    Check other cross-columns for the sum of the first n elements.  Each time
      double the sum and look for the location of that value in the Duplex
      Triangle.

In general, the sum of the first n + 1 elements of cross-column c of the Duplex
Triangle is equal to one-half the the n-th element in cross-column c + 1. That is,

**N(c, 0) + N(c + 1, 1) + N(c + 2, 2) + N(c + 3, 3) + . . . + N(c + n, n)**
                                    **= (½) • N(c + n + 1, n)**       (2)

6.     Let c = 3 and n = 4 in equation (2).  Find the sum of the first 5
       elements of cross-column c.  If you double the sum, in what row and
       column do you find this value?

A property of Pascal's Triangle, sometimes called the "hockey stick" property, is similar to equation (2).  Namely, the sum of the first n + 1 elements of cross-column c of Pascal's Triangle is equal to the n-th element in cross-column c + 1.  That is,

$$C(c, 0) + C(c + 1, 1) + C(c + 2, 2) + C(c + 3, 3) + \ldots + C(c + n, n) = C(c + n + 1, n) \qquad (3)$$

Equation (3) is illustrated below in Pascal's Triangle for c = 2 and n =4.

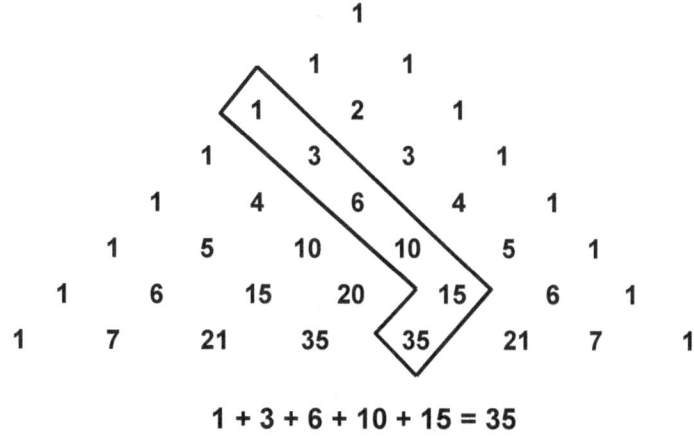

$$1 + 3 + 6 + 10 + 15 = 35$$

**Figure 4.  Illustrating the "hockey-stick" property of Pascal's Triangle.**

Because of the symmetry in Pascal's Triangle, the hockey stick property holds for columns, as well as, cross-columns.

7.     Let c = 1 and n = 5 in equation (3) above.  Find the sum of the first 6
       elements of cross-column c.  In what row and column do you find this
       value? Sketch in the corresponding "hockey-stick" .

8.     In each of the four Pascal's Triangles that follow, add all the numbers
       in the top group, then add all the numbers enclosed in the bottom
       group.  Compare the two sums.  What do you find?

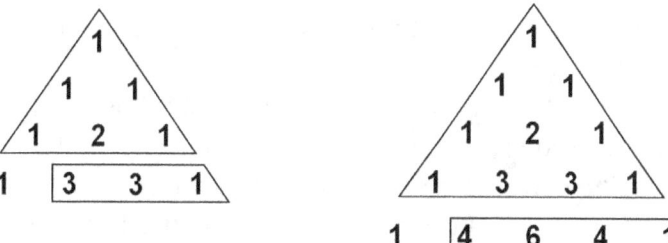

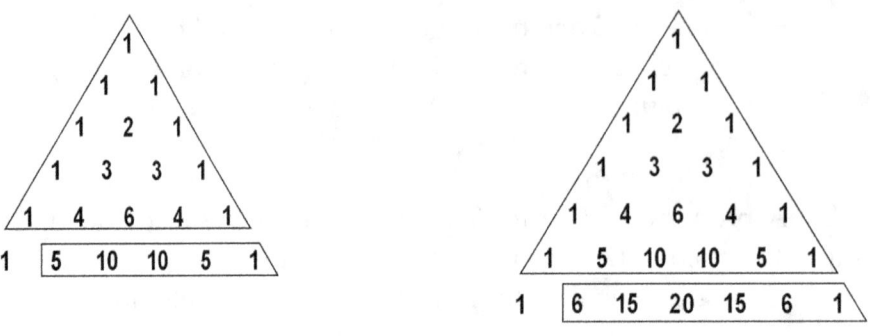

The hockey stick property can verify what we discovered above, that is, the sum of all the elements in Pascal's Triangle up through the nth row is equal to $2^{n+1} - 1$.

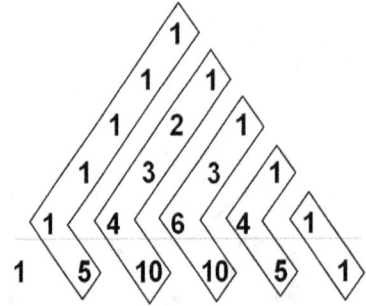

**Figure 5.  The sum of all the elements up through the $4^{th}$ row of Pascal's Triangle is equal to 1 less than the sum of the elements in the $5^{th}$ row, as demonstrated by the hockey-stick property.**

Since the sum of the elements in each row of Pascal's Triangle is $2^n$, we have the following formula corresponding to the figure above:

$$2^0 + 2^1 + 2^2 + 2^3 + 2^4 = 2^5 - 1$$

In general, the formula for adding all the numbers in Pascal's Triangle up through the nth row is

$$2^0 + 2^1 + 2^2 + \cdots + 2^{n-1} + 2^n = 2^{n+1} - 1 \tag{4}$$

9.    Use the formula (4) to find the sum of the powers of 2 from the $0^{th}$ through the $10^{th}$.

10.    Devise a plan to find the sum of all the numbers in Pascal's Triangle from row 8 through row 12.

11.    In each of the Duplex Triangles that follow, add all the numbers in the top group, then add all the numbers enclosed in the bottom group. Show in each case that sum of the numbers in the top group is equal to one-half the sum of the numbers in the bottom group.

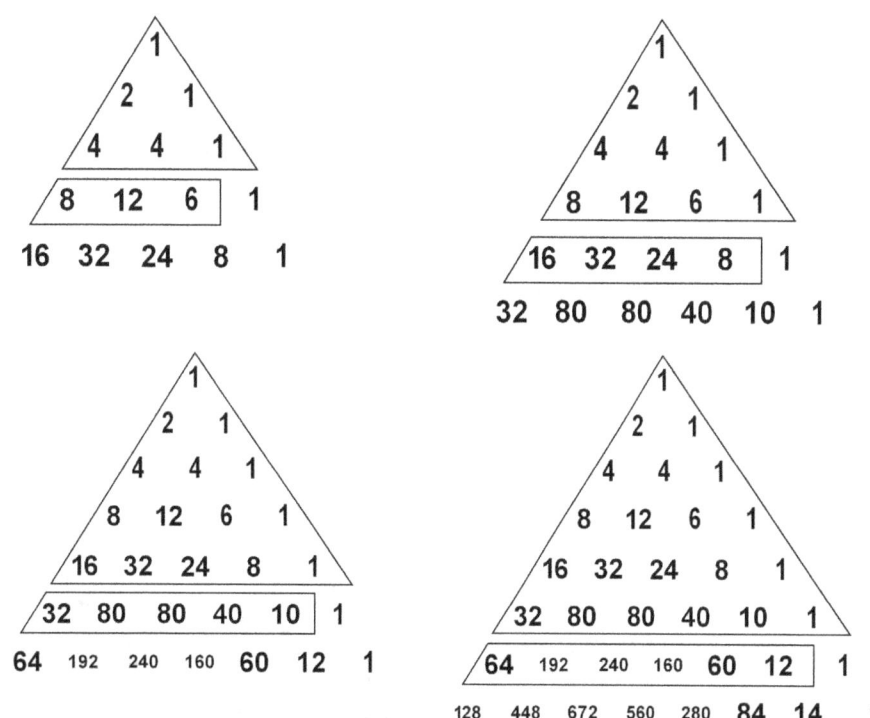

12. In each of the Duplex Triangles that follow, add all the numbers in the top group, then add all the numbers enclosed in the bottom group. Show in each case that sum of the numbers in the top group is equal to one-half the sum of the numbers in the bottom group.

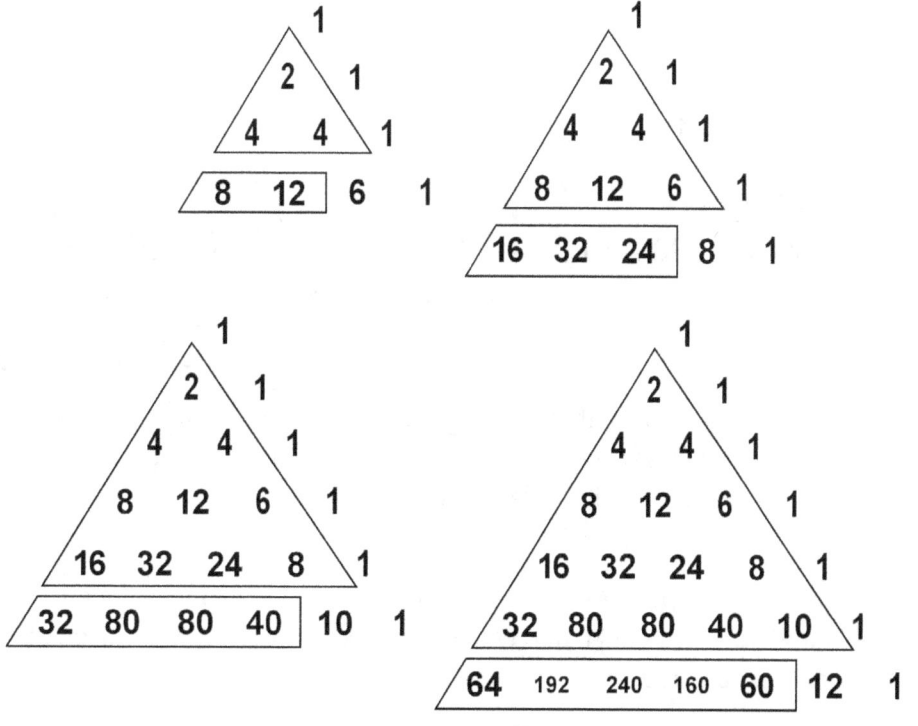

13.    In each of the Duplex Triangles that follow, add all the numbers in the top group, then add all the numbers enclosed in the bottom group. Show in each case that sum of the numbers in the top group is equal to one-half the sum of the numbers in the bottom group.

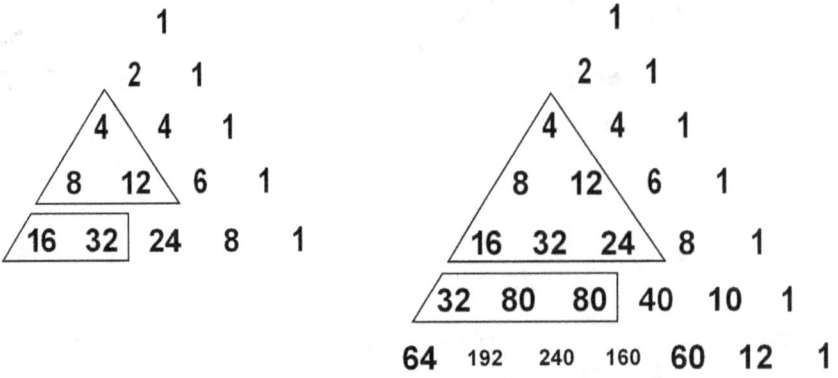

14.    Experiment with another group of numbers in a manner similar to the groups shown in Exercise 13, also starting with the element 4 at the apex the triangular group.

15.    In each of the Duplex Triangles that follow, add all the numbers in the top group, then add all the numbers enclosed in the bottom group. Show in each case that sum of the numbers in the top group is equal to one-half the sum of the numbers in the bottom group.

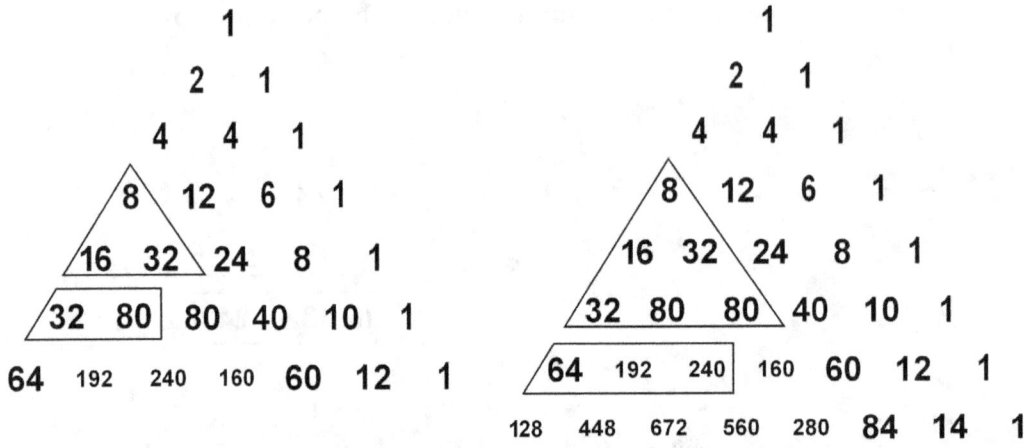

16.    Experiment with another group of numbers in a manner similar to the groups shown in Exercise 14, also starting with the element 8 at the apex the triangular group.

Expanding the property expressed in equation (2) to all of the cross-columns of the Duplex Triangle., illustrates what we discovered, that is, the sum of all the elements in Duplex Triangle up through the nth row is equal to

$$(1/2)(3^{n+1} - 1).$$

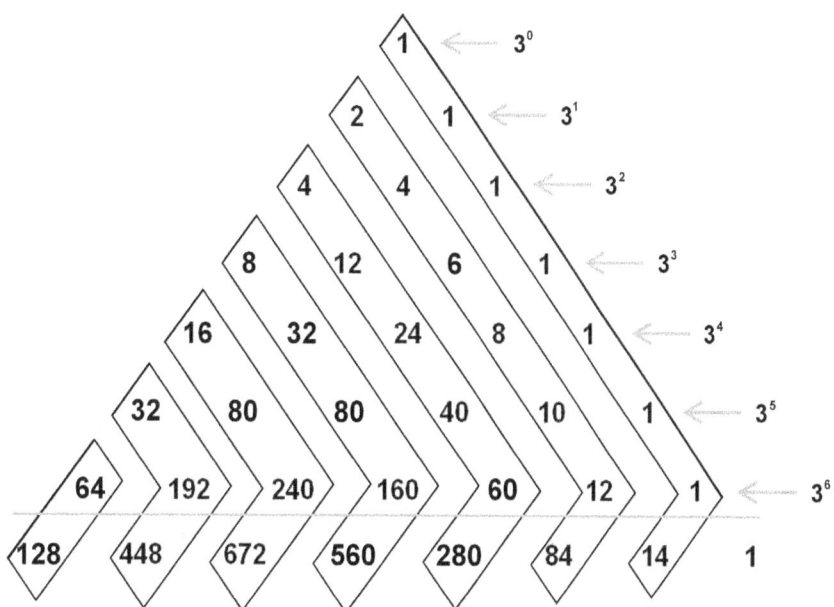

**Figure 6.** In the Duplex Triangle shown above, the sum of each cross-column element up through the 6$^{th}$ row is equal to ½ the corresponding element in the 7$^{th}$ row.

The sum of all of the elements in the Duplex Triangle up through the 6$^{th}$ row is equal to ½ the sum of the elements in the 7$^{th}$ row minus 1.     Each row sum is equal to a power of 3. Therefore,

$$3^0 + 3^1 + 3^2 + 3^3 + 3^4 + 3^5 + 3^6 = (1/2)(3^7 - 1)$$

In general, we have a formula for the sum of the powers of 3 as follows:

$$3^0 + 3^1 + 3^2 + \ldots + 3^n = (1/2)(3^{n+1} - 1) \qquad (5)$$

15.     Use the formula (5) to find the sum of the powers of 3 from the 0$^{th}$ through the 10$^{th}$.

16.     Devise a plan to find the sum of all the numbers in the Duplex Triangle from row 6 through row 10.

Figure 7 shows the development of an interesting sequence of numbers.

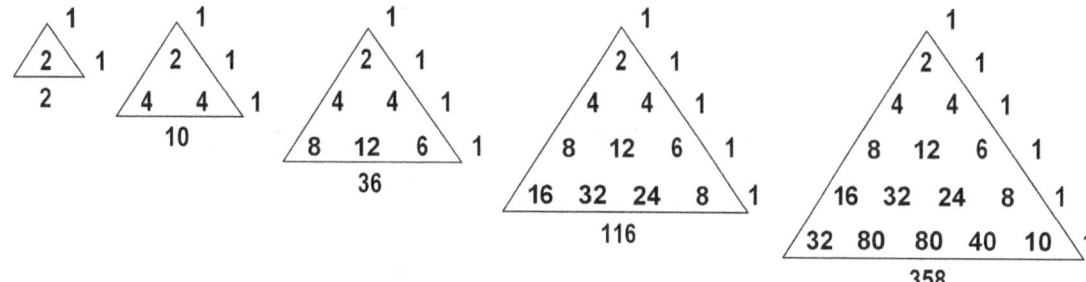

**Figure 7. Sums of all the elements of the Duplex Triangle (without the trailing 1's) through rows 1, 2, . . . , 5.**

The sequence of sums developed in Fig. 7 is 2, 10, 36, 116, 358.  We begin by multiplying each element in the sequence by ½, yielding the sequence

**1, 5, 18, 58, 179**.

---

**17.**    Find the next number in the sequence **1, 5, 18, 58, 179**, . . .

---

Next we form the **difference sequence** from our sequence as follows:

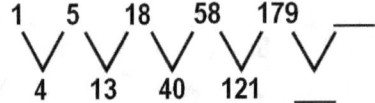

Each entry on the bottom row is the difference of the two entries above it.   The new sequence begins with a 1.   Next form the **difference sequence** from our new sequence as follows:

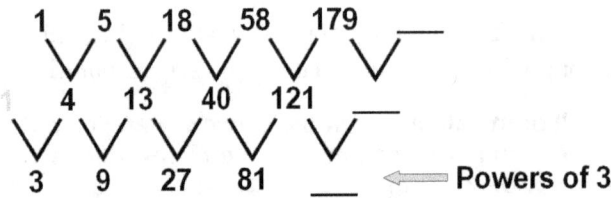

This newly formed sequence is the sequence of the powers of 3.

---

**18.**    Fill in the missing entries in the difference sequences shown above.

**19.**    Form the sequence of sums of the enclosed elements of the Duplex Triangles shown below, then divide each element in the sequence by 4. Next form the **difference sequence** from your sequence, and precede it with a 1.  Continue forming difference sequences from your newly formed sequences until you find the  sequence of the powers of 3.

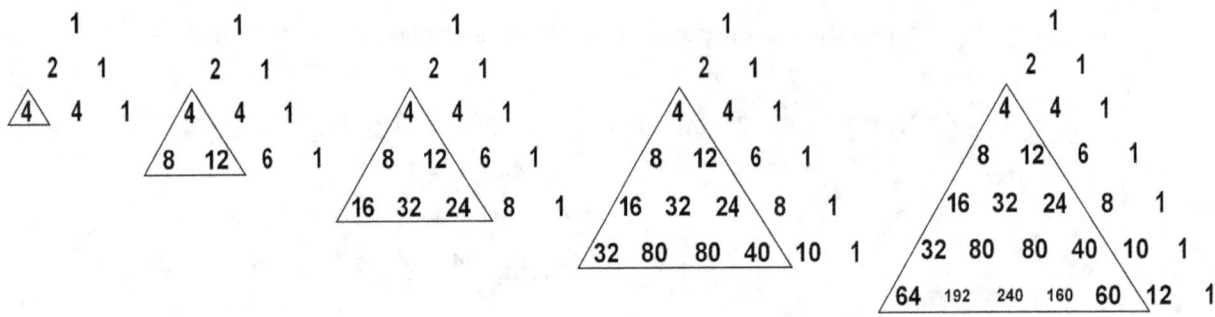

---

## 3.3 Number Patterns Related to the Columns of the Duplex Triangle

The "hockey-stick" property of Pascal's Triangle is equally valid for columns and cross-columns because of the symmetry of Pascal's Triangle. In the previous section we found that there is a similar property for the cross-columns of the Duplex Triangle. For the columns of the Duplex Triangle, however, we find a somewhat different, related property. This property involves two columns and is demonstrated below. First, lets review the layout of the columns of the Duplex Triangle.

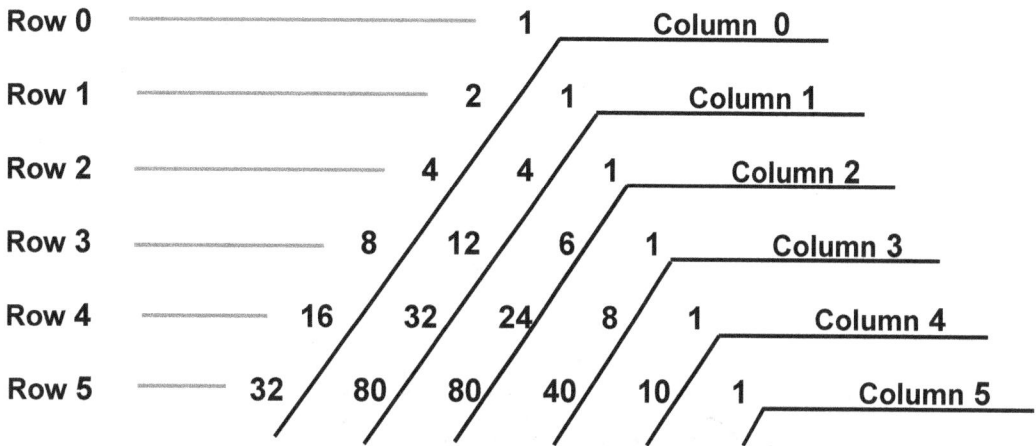

Figure 8. Rows and columns of the Duplex Triangle.

If we add all of the elements in two adjacent columns down to the nth row, the sum is equal to the element in the (n + 1)th row and the column equal to the column number of the second of the two adjacent columns. For example, the sum of the elements in the 2nd and 3rd columns down to the 5th row is equal to the element in the 6th row and 3rd column (Fig. 9).

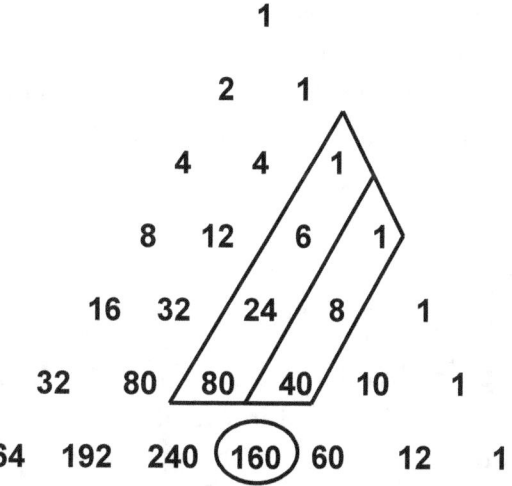

Figure 9. The sum of the elements in the columns above the element 160 (encircled) is equal to that element, 160.

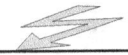

**Section 3.3**

1.    Verify that the sum of the elements in the 1st and 2nd columns down to the 4th row is equal to the element in the 5th row and 2nd column.

2.    Verify that the sum of the elements in the 3rd and 4th columns down to the 5th row is equal to the element in the 6th row and 4th column.

3.    Experiment with other columns and rows and verify the results.

4.    **(a)**  In the figure below add all of the elements of the Duplex Triangle down to the 5th row.  Show that the sum is equal to the sum of every other element in the 6th row (the encircled elements).

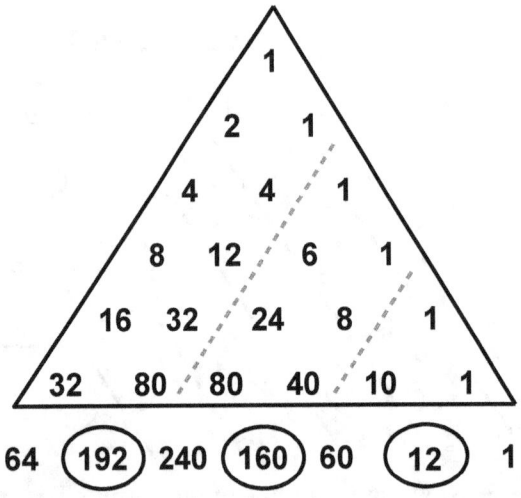

**(b)** Show that the sum  (192 + 160 + 12) in the 6th row is equal to $(3^6 - 1)/2$.

5.    **(a)**  Add all of the elements of the Duplex Triangle down to the 4th row. Show that the sum is equal to the sum of every other element in the 5th row, starting with column 1.

**(b)** Show that the sum  of every other element in the 5th row, starting with column 1, is equal to $(3^5 - 1)/2$.

**(c)** Show that 1 less than the sum  of every other element in the 5th row, starting with column 0,  is equal to   $(3^5 - 1)/2$ .

## 3.4  Number Patterns of the Duplex Triangle Involving String Products and String Dot Products

String and string dot products were discussed in Chapter 2.  Another operation involving strings is called the **scalar product**. The result of this operation is a new string where each element of the operand string is multiplied by a number, called a scalar.  For example,

**4 (1, 3, 6, 10) = (4, 12, 24, 40).**

There are two things to notice in this example. First, the elements of the resulting string, the scalar product, are the first 4 elements of the cross-column 2 of the Duplex Triangle. Second, the elements of the operand string are the first 4 elements of column 2 of Pascal's Triangle, also known as the triangular numbers.

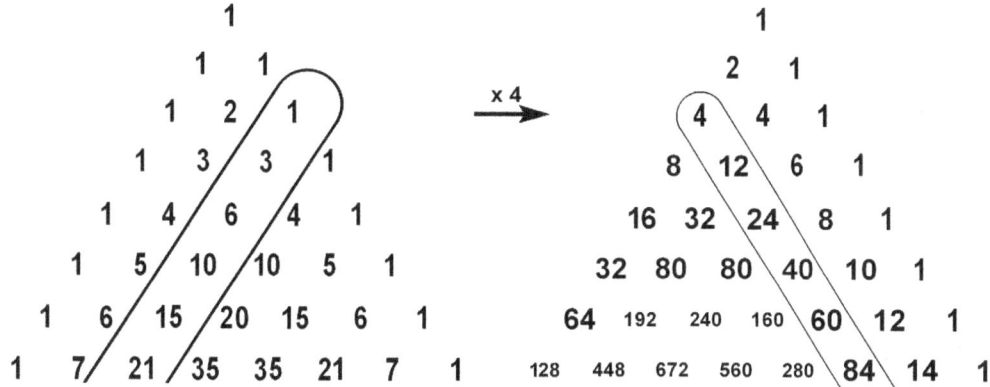

Figure 10. Left: Pascal's Triangle, Column 2 (triangular numbers).
Right: Duplex Triangle, Cross-column 2.
4 times a given triangular number is equal to a corresponding element in the
Duplex Triangle, Cross-column 2.

 **Section 3.4**

1. Find the scalar product of 4 and the first 5 elements of column 2 of Pascal's Triangle, that is, the first 5 triangular numbers. Compare to the first 5 elements of the cross-column 2 of the Duplex Triangle.

2. Find the scalar product of 2 and the first 5 elements of column 1 of Pascal's Triangle, that is, the first 5 natural numbers. Compare to the first 5 elements of the cross-column 1 of the Duplex Triangle.

3. Find the scalar product of 8 and the first 5 elements of column 3 of Pascal's Triangle, that is, the first 5 tetrahedral numbers. Compare to the first 5 elements of the cross-column 3 of the Duplex Triangle.

The 3 exercises above are specific examples of a more general property, namely,

**the elements of the cross-column c of the Duplex Triangle are equal to the scalar product of $2^c$ and the elements of column c of Pascal's Triangle.**

4. Find the scalar product of $2^4$ and the first few elements of column 4 of Pascal's Triangle. Compare to the first few elements of the cross-column 4 of the Duplex Triangle.

In Sec. 2.5 we discussed the extensions of the Pascal's Triangle Rule and the Duplex Triangle Rule. In general, if we form the dot product of the nth and mth rows, we get the element in th (n + m)th row and the cth column, which is the the column position of the right-most entry of the string representing the longer row. Of course, to get proper string dot products we have to extend the shorter row with enough zeros to match the length of the longer row. And with the Duplex Triangle we also have to reverse the string representing the shorter row before forming the dot product.

In the next set of exercises we will see that string dot products can also be formed from strings made up of elements representing only partial rows. For example, form the string dot product of the last 3 three elements of row 3 of the Duplex Triangle with the last three elements of the $4^{th}$ row (we reverse the string representing the $3^{rd}$ row).

$$(1, 6, 12) \bullet (24, 8, 1) = 24 + 48 + 12 = 84 = N(7, 5)$$

The result, 84, is found in the $7^{th}$ row (3 + 4 = 7) and the $5^{th}$ column, N(7, 5) = 84. The column number (5) of the result is determined by adding the Duplex Triangle column positions of the elements in the last positions of the two operand strings. In our example those elements are 12 in the first string and 1 in the second string and their column positions in the Duplex Triangle are 1 and 4, respectively. Adding these numbers gives 5, the column position of the result, 84. See Fig. 11.

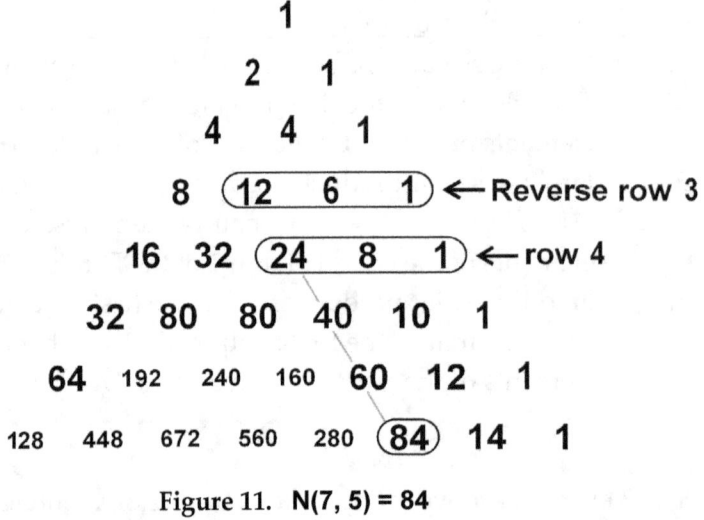

Figure 11.  N(7, 5) = 84

Another example, form the string dot product of the last two elements of the $2^{nd}$ row of the Duplex Triangle with the last two elements of the $4^{th}$ row.

$$(1, 4) \bullet (8, 1) = 8 + 4 = 12$$

This time the result, 12, is found in the $6^{th}$ row (2 + 4 = 6), and in the $5^{th}$ column, N(6, 5) = 12. Column 5 of the result is determined by the element 4 in the first operand string, which is in column position 1 of the Duplex Triangle and the element 1 in the second string which is found in the column position 4 in the $4^{th}$ row of the Duplex Triangle. The sum of these column positions is 1 + 4 = 5.

5.     Compute this string dot product:    $(1, 6, 12) \bullet (12, 6, 1) = ?$
        In what row and column can the result be found in the Duplex Triangle?

6.     Compute this string dot product:    $(1, 4) \bullet (6, 1) = ?$
        In what row and column can the result be found in the Duplex Triangle?

7.     Compute this string dot product:    $(1, 4, 4) \bullet (24, 8, 1) = ?$
        In what row and column can the result be found in the Duplex Triangle?

8.     Compute this string dot product:    $(1, 6, 12, 8) \bullet (32, 24, 8, 1) = ?$
        In what row and column can the result be found in the Duplex Triangle?

9.     Compute this string dot product:    $(1, 8, 24) \bullet (24, 8, 1) = ?$
        In what row and column can the result be found in the Duplex Triangle?

10.    Compute this string dot product:    $(1, 6, 12) \bullet (40, 10, 1) = ?$
        In what row and column can the result be found in the Duplex Triangle?

---

At the beginning of this Chapter we reviewed two formulas for the element in the Duplex Triangle at row d and column k. The first formula was the explicit formula

$$\mathbf{N(d, k) = C(d, d - k) \cdot 2^{d-k} = C(d, k) \cdot 2^{d-k},}$$

where $C(d, k)$ is the binomial coefficient, the element in row d, column k of Pascal's Triangle. The second formula was the Duplex Triangle Rule, a recursive rule for the element in the d-th row and k-th column of the Duplex Triangle. It is similar to Pascal's Rule for Pascal's Triangle, except here the second element in the right-hand member is multiplied by 2,

$$\mathbf{N(d, k) = N(d - 1, k - 1) + 2 \cdot N(d - 1, k).}$$

Now we present another formula for $N(d, k)$ in the form of a string dot product. The strings in this product come from Pascal's Triangle. We will form the product of row d of Pascal's Triangle and column k of Pascal's Triangle.

$$\mathbf{N(d, k) = (row\ d) \bullet (column\ k),}\ \text{row and column from Pascal's Triangle}$$

For example, using row 4 and column 2 of Pascal's Triangle we have

$$\mathbf{N(4, 2) = (1, 4, 6, 4, 1) \bullet (0, 0, 1, 3, 6)}$$
$$\mathbf{= 0 + 0 + 6 + 12 + 6}$$
$$\mathbf{= 24}$$

Notice the zeros in the second string. This is because the last element of the second string is required to be in the $4^{th}$ row, the same as the row string used in the product. See Figure 12.

The row number d, in the expression $N(d, k)$, refers to the dimension of a duplex figure and the column number k refers to the dimension of the duplex sub-elements of that duplex figure. Thus, for example, the value $N(4, 2) = 24$ indicates that in a 4-D duplex there are 24 2-D sub-elements, that is, 24 square faces.

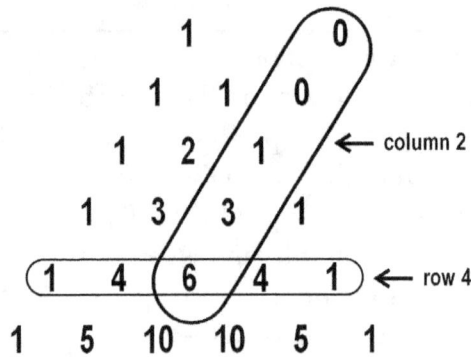

**Figure 12.  The string dot product of a row and a column of Pascal's Triangle
equals an element in the Duplex Triangle.**
**N(4, 2) =  (1, 4, 6, 4, 1) • (0, 0, 1, 3, 6) = 24**

---

11.    Compute this string dot product:    (1, 3, 3, 1) • (0, 1, 2, 3) = ?
       In what row and column can the result be found in the Duplex Triangle?

12.    Compute this string dot product:    (1, 5, 10, 10, 5, 1) • (0, 0, 1, 3, 6, 10) = ?
       In what row and column can the result be found in the Duplex Triangle?

13.    Compute this string dot product:    The 4th row of Pascal's Triangle and
       the first column of Pascal's Triangle.
       In what row and column can the result be found in the Duplex Triangle?

---

## 3.5  Vandermonde's Identity for the Duplex Triangle

Vandermonde's Identity involves the combinatorials in Pascal's Triangle and has
applications in the field of combinatorics.   Remarkably, there is a similar property that
holds for the Duplex Triangle.   First, we will review Vandermonde's Identity for Pascal's
Triangle.

Essentially, Vandermonde's Identity says that if we choose a row of **Pascal's Triangle**,
call it the rth row, and split r into two parts, m and n,  so that $m + n = r$, then an
element anywhere in that row, say in column position c, is equal  to  the  string  dot
product of the string of elements in row m from column 0 to column c and the string
of elements in row n from column c back to column 0.

Here is an example using this description.  Select row 9 of Pascal's Triangle, r = 9,
and split 9 into two parts.  We will arbitrarily  split it  into the parts 6 and 3,  so m = 6   and
n = 3  and  $m + n = r = 9$.  Now choose a column position in row 9, we will arbitrarily choose
c = 5.  We now have the element $C(r, c) = C(9, 5) = 9! / (5! • 4!) = 126$.  The identity states that
this element is equal to the string dot product of the string in the m th row ($6^{th}$ row) from
column 0  to c = 5, and the string in the nth row ( $3^{rd}$ row) from column c = 5 down to 0.
That is,

$$C(9, 5) = (1, 6, 15, 20, 15, 6) \cdot (0, 0, 1, 3, 3, 1)$$
$$= 1\cdot 0 \ + \ 6\cdot 0 \ + \ 15\cdot 1 \ + \ 20\cdot 3 \ + \ 15\cdot 3 \ + \ 6\cdot 1$$
$$= 0 + 0 + 15 + 60 + 45 + 6$$
$$= 126$$

Since the $3^{rd}$ row of Pascal's Triangle does not have a $5^{th}$ nor a $4^{th}$ column element, 0's are filled into the string to make the number of positions match the other string. The position of the rows and elements are seen in Fig. 13.

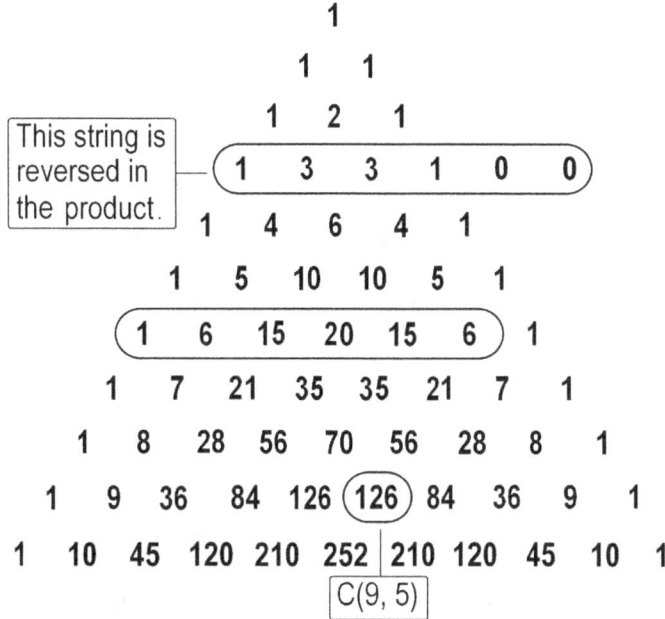

Figure 13. **Illustration of Vandermonde's Identity for Pascal's Triangle**

$$C(9, 5) = (1, 6, 15, 20, 15, 6) \cdot (0, 0, 1, 3, 3, 1)$$
$$= 0 + 0 + 15 + 60 + 45 + 6$$
$$= 126$$

The same arrangement and configuration of elements holds for the Duplex Triangle.

---

**Vandermonde's Identity for the Duplex Triangle**

Choose a row of the Duplex Triangle, call it the rth row, and split r into two parts, m and n, so that m + n = r, then an element anywhere in that row, say in column position c, is equal to the string dot product of the string of elements in row m from column 0 to column c and the string of elements in row n from column c back to column 0.

---

For example, select row 7 of the Duplex Triangle, r = 7, and split it into two parts. We will arbitrarily split it into the parts 4 and 3, so m = 4 and n = 3 and m + n = r = 7. Now choose a column position in row 7, we will arbitrarily choose c = 3. We now have the element $N(r, c) = N(7, 3) = C(7, 7 - 3)\cdot 2^{7-3} = 35\cdot 16 = 560$. The identity states that this element is equal to the string dot product of the string in the m th row ($4^{th}$ row) of the Duplex Triangle from column 0 to c = 3, and the string in the nth row ($3^{rd}$ row) from

column c = 3 down to 0. That is,

$$N(7, 3) = (16, 32, 24, 8) \bullet (1, 6, 12, 8)$$
$$= 16 \bullet 1 + 32 \bullet 6 + 24 \bullet 12 + 8 \bullet 8$$
$$= 16 + 192 + 288 + 64$$
$$= 560$$

The position of the rows and elements are seen in Fig. 14.

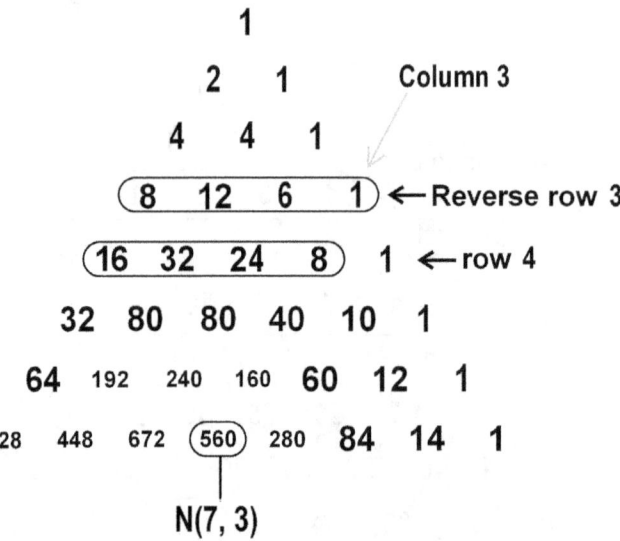

Figure 14.  Illustration of Vandermonde's Identity for the Duplex Triangle
$$N(7, 3) = (16, 32, 24, 8) \bullet (1, 6, 12, 8)$$
$$= 16 \bullet 1 + 32 \bullet 6 + 24 \bullet 12 + 8 \bullet 8$$
$$= 16 + 192 + 288 + 64$$
$$= 560$$

If the smaller row of the Duplex Triangle did not have a enough column elements to match the length of the string from the longer row, then 0's are prepended into the string of elements of the shorter row to make the number of positions equal in the two strings.   For example,  choose row 5  and let m = 3, n = 2, and choose c = 4.   Then,

$$N(5, 4) = (8, 12, 6, 1, 0) \bullet (0, 0, 1, 4, 4)$$
$$= 8 \bullet 0 + 12 \bullet 0 + 6 \bullet 1 + 1 \bullet 4 + 0 \bullet 4$$
$$= 0 + 0 + 6 + 4 + 0 = 10$$

In this example, c was greater than the last column position of the longer string (corresponding to row 3), so it was necessary to append a zero to that string.  Therefore, we prepended two zeros to the string corresponding to the reverse of row 2 to match the length of the first string.

 **Section  3.5**

1.       Select row 7 of Pascal's Triangle, r = 7,  and split it into two parts so m = 4 and  n = 3  and  m + n = r = 7.  Let the column position in row 7 equal 3, c = 3.  Form the two strings described in Vandermonde's Identity and show that the string dot product equals C(7, 3) = 35.

2.     Select row 6 of the Duplex Triangle, $r = 6$, and split it into two parts so $m = 4$ and $n = 2$ and $m + n = r = 6$. Let the column position in row 6 equal 3, $c = 3$. Form the two strings described in Vandermonde's Identity for the Duplex Triangle and show that the string dot product equals $N(6, 3) = 160$.

3.     Select row 6 of the Duplex Triangle, $r = 6$, and split it into two parts so $m = 3$ and $n = 3$ and $m + n = r = 6$. Let the column position in row 6 equal 3, $c = 3$. Form the two strings described in Vandermonde's Identity for the Duplex Triangle and show that the string dot product equals $N(6, 3) = 160$.

4.     Select row 6 of the Duplex Triangle, $r = 6$, and split it into two parts so $m = 5$ and $n = 1$ and $m + n = r = 6$. Let the column position in row 6 equal 3, $c = 3$. Form the two strings described in Vandermonde's Identity for the Duplex Triangle and show that the string dot product equals $N(6, 3) = 160$.

5.     Select row 6 of the Duplex Triangle, $r = 6$, and split it into two parts so $m = 3$ and $n = 3$ and $m + n = r = 6$. Let the column position in row 6 equal 5, $c = 5$. Form the two strings described in Vandermonde's Identity for the Duplex Triangle and show that the string dot product equals $N(6, 5) = 12$.

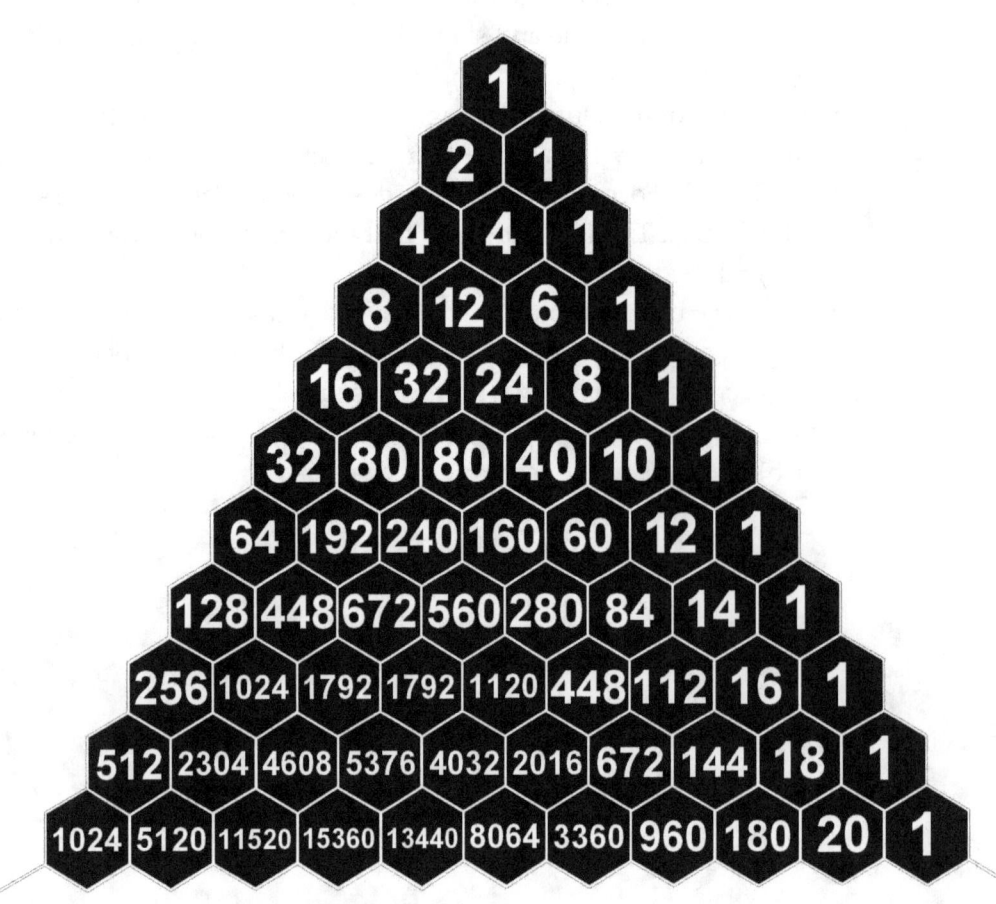

**Duplex Triangle**

# Diagonal Patterns Within the Duplex Triangle
## - The Pell Numbers

## 4.1  The Pell and Fibonacci Numbers

The infinite sequence of integers

$$0, 1, 2, 5, 12, 29, 70, 169, 408, 985, \ldots, P_n, \ldots$$

where **n = 0, 1, 2, 3,  . . .**, and $P_n$ represents the nth number in the sequence, is named the Pell sequence, or the elements in the sequence are called Pell Numbers  (John Pell, English mathematician, 1611-1685).  The significance of the Pell numbers as they relate to the Duplex Triangle, is that the sums of the elements  on the diagonals of the Duplex Triangle are the Pell numbers, just as the Fibonacci numbers are sums of the diagonal elements of Pascal's Triangle.

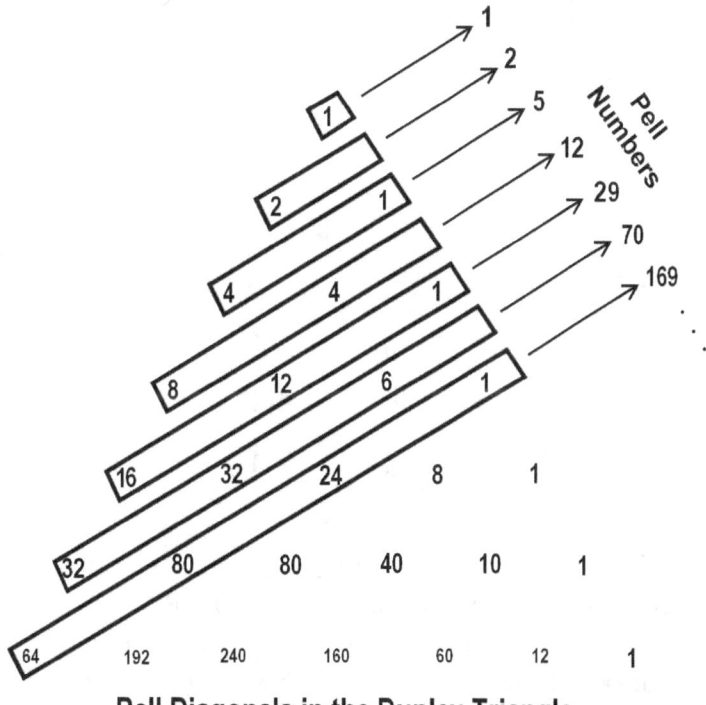

**Pell Diagonals in the Duplex Triangle**

On the 5th diagonal , for example, we have $P_5 = 16 + 12 + 1 = 29$.  See the figure above.

Like the Fibonacci sequence, a formula for the Pell sequence is a recursive formula where each element is based on the previous two elements in the sequence. Recall the Fibonacci sequence and let $F_n$ represent the nth Fibonacci number in the sequence, then $F_{n-1}$ and $F_{n-2}$ represent two previous consecutive Fibonacci numbers in the sequence. The recursive formula, called the rule Fibonacci Rule is

$$F_n = 1 \cdot F_{n-1} + 1 \cdot F_{n-2}$$

where $F_1 = 1$ and $F_2 = 1$ and $n = 3, 4, 5, \ldots,$

Yielding the infinite sequence $1, 1, 2, 3, 5, 8, 13, 21, 34, \ldots, F_n, \ldots$ .

Unlike the Fibonacci sequence, an element in the Pell sequence is equal to 2 times the previous element plus the element just prior to that. The following is the recursive Pell rule.

$$P_n = 2 \cdot P_{n-1} + P_{n-2}$$

for $n = 2, 3, \ldots$ and $P_0 = 0$ and $P_1 = 1.$

For example, $P_5 = 2 \cdot P_4 + P_3$, or equivalently, $29 = 2 \cdot 12 + 5.$

*The Pell sequence is to the Duplex Triangle*
*as the Fibonacci sequence is to Pascal's Triangle.*

The Fibonacci rule is based on row 1 of Pascal's Triangle, in that it is the <u>string dot product</u> of that row with two previous consecutive elements in the Fibonacci sequence. This gives us the following alternative version of the Fibonacci Rule:

$$F_n = (1, 1) \bullet (F_{n-1}, F_{n-2}) = 1 \bullet F_{n-1} + 1 \bullet F_{n-2}.$$

In general, we have the following recursive formulas, which are the Pascal extensions of the Fibonacci Rule.

$$F_n = 1 \cdot F_n$$

$$= 1 \cdot F_{n-1} + 1 \cdot F_{n-2} \tag{1}$$

$$= 1 \cdot F_{n-2} + 2 \cdot F_{n-3} + 1 \cdot F_{n-4} \tag{2}$$

$$= 1 \cdot F_{n-3} + 3 \cdot F_{n-4} + 3 \cdot F_{n-5} + 1 \cdot F_{n-6} \tag{3}$$

$$= 1 \cdot F_{n-4} + 4 \cdot F_{n-5} + 6 \cdot F_{n-6} + 4 \cdot F_{n-7} + 1 \cdot F_{n-8} \tag{4}$$

$$= 1 \cdot F_{n-5} + 5 \cdot F_{n-6} + 10 \cdot F_{n-7} + 10 \cdot F_{n-8} + 5 \cdot F_{n-9} + 1 \cdot F_{n-10} \tag{5}$$

etc.

Formulas (1) thru (5) represent the <u>string dot products</u> of the rows of Pascal's Triangle with consecutive elements of the Fibonacci sequence. For instance, if we use row 3 of Pascal's Triangle in the product, then the consecutive sequence of Fibonacci numbers begins with $F_{n-3}$ and works backward for three more Fibonacci numbers, or a total of four consecutive Fibonacci numbers (in this case it is assumed that n is at least 6):

$$F_n = (1, 3, 3, 1) \bullet (F_{n-3}, F_{n-4}, F_{n-5}, F_{n-6}) = 1 \cdot F_{n-3} + 3 \cdot F_{n-4} + 3 \cdot F_{n-5} + 1 \cdot F_{n-6}.$$

Working with the Duplex Triangle and the Pell sequence, we have a similar set of recursive formulas for the nth Pell number. The recursive rule for the Pell numbers is formed from the <u>string dot product</u> of row 1 of the Duplex Triangle and two previous consecutive elements of the Pell sequence, giving us

$$P_n = (2, 1) \bullet (P_{n-1}, P_{n-2}) = 2 \cdot P_{n-1} + P_{n-2}$$

In general, we have the following recursive formulas, which are the Duplex extensions of the recursive Pell Rule.

$$P_n = \; 1 \cdot P_n$$

$$= \; 2 \cdot P_{n-1} + \; 1 \cdot P_{n-2} \tag{6}$$

$$= \; 4 \cdot P_{n-2} + \; 4 \cdot P_{n-3} + \; 1 \cdot P_{n-4} \tag{7}$$

$$= \; 8 \cdot P_{n-3} + \; 12 \cdot P_{n-4} + \; 6 \cdot P_{n-5} + \; 1 \cdot P_{n-6} \tag{8}$$

$$= 16 \cdot P_{n-4} + 32 \cdot P_{n-5} + 24 \cdot P_{n-6} + \; 8 \cdot P_{n-7} + \; 1 \cdot P_{n-8} \tag{9}$$

$$= 32 \cdot P_{n-5} + 80 \cdot P_{n-6} + 80 \cdot P_{n-7} + 40 \cdot P_{n-8} + 10 \cdot P_{n-9} + 1 \cdot P_{n-10} \tag{10}$$

etc.

Formulas (6) thru (10) represent the <u>string dot products</u> of the rows of the Duplex Triangle with previous consecutive elements of the Pell sequence. For instance, if we use row 2 of the Duplex Triangle in the product, then the consecutive sequence of Pell numbers begins with $P_{n-2}$ and works backward for two more Pell numbers, a total of three consecutive Pell numbers. This assumes that n is at least equal to 4.

$$P_n = (4, 4, 1) \bullet (P_{n-2}, P_{n-3}, P_{n-4}) = 4 \cdot P_{n-2} + 4 \cdot P_{n-3} + 1 \cdot P_{n-4}$$

**Example:** Verify formula (7) for $n = 6$.
**Solution:** $P_6 = 70$ and

$$4 \cdot P_{n-2} + 4 \cdot P_{n-3} + 1 \cdot P_{n-4} = 4 \cdot P_4 + 4 \cdot P_3 + 1 \cdot P_2$$
$$= 4 \cdot 12 + 4 \cdot 5 + 1 \cdot 2$$
$$= \; 48 \; + \; 20 \; + \; 2$$
$$= \; 70$$

We can also prove formula (7) by using formula (6), the given recursive rule for the Pell sequence.

Given Formula (6):  $P_n = \; 2 \cdot P_{n-1} + \; 1 \cdot P_{n-2}$
Use the recursive rule and replace $P_{n-1}$ and $P_{n-2}$ with
$2 \cdot P_{n-2} + 1 \cdot P_{n-3}$ and $2 \cdot P_{n-3} + 1 \cdot P_{n-4}$, respectively.

Therefore,

$$P_n = 2 \cdot P_{n-1} + 1 \cdot P_{n-2}$$
$$= 2 \cdot [2 \cdot P_{n-2} + 1 \cdot P_{n-3}] + 1 \cdot [2 \cdot P_{n-3} + 1 \cdot P_{n-4}]$$
$$= 4 \cdot P_{n-2} + 2 \cdot P_{n-3} + 2 \cdot P_{n-3} + 1 \cdot P_{n-4}$$
$$= 4 \cdot P_{n-2} + 4 \cdot P_{n-3} + 1 \cdot P_{n-4} \qquad \blacksquare$$

### Section 4.1

1.    Verify formula (2) for **(a)** $n = 5$ and **(b)** $n = 6$.
2.    Verify formula (3) for **(a)** $n = 7$ and **(b)** $n = 8$.
3.    Prove formula (2) by using formula (1), the recursive rule for the Fibonacci sequence.
4.    Prove formula (3) by using formula (1), the recursive rule for the Fibonacci sequence.
5.    Verify formula (7) for $n = 8$.
6.    Verify formula (8) for $n = 8$.
7.    Prove formula (8) by using formula (6), the recursive rule for the Pell sequence.
8.    Write the string dot product using row 6 of the Duplex Triangle and the the appropriate sequence of consecutive Pell numbers.
9.    Show that $P_n = P_{n-1} + 3P_{n-2} + P_{n-3}$.

The Pell recursive formula also fits nicely into another pattern of recursive forumlas for $P_n$.

$$P_n = 2 \cdot P_{n-1} + 1 \cdot P_{n-2} = P_2 \cdot P_{n-1} + P_1 \cdot P_{n-2}, \ n \geq 2 \qquad (11)$$
$$= 5 \cdot P_{n-2} + 2 \cdot P_{n-3} = P_3 \cdot P_{n-2} + P_2 \cdot P_{n-3}, \ n \geq 3 \qquad (12)$$
$$= 12 \cdot P_{n-3} + 5 \cdot P_{n-4} = P_4 \cdot P_{n-3} + P_3 \cdot P_{n-4}, \ n \geq 4 \qquad (13)$$
$$= 29 \cdot P_{n-4} + 12 \cdot P_{n-5} = P_5 \cdot P_{n-4} + P_4 \cdot P_{n-5}, \ n \geq 5 \qquad (14)$$
$$= 70 \cdot P_{n-5} + 29 \cdot P_{n-6} = P_6 \cdot P_{n-5} + P_5 \cdot P_{n-6}, \ n \geq 6 \qquad (15)$$

etc.

**Example:** Verify formula (13) for $n = 6$.
**Solution:** $P_6 = 70$ and

$$P_4 \cdot P_{n-3} + P_3 \cdot P_{n-4} = 12 \cdot P_{n-3} + 5 \cdot P_{n-4}$$
$$= 12 \cdot P_3 + 5 \cdot P_2$$
$$= 12 \cdot 5 + 5 \cdot 2$$
$$= 60 + 10$$
$$= 70$$

10.     Verify formula (13) for n = 5.
11.     Verify formula (14) for n = 7.
12.     Verify formula (15) for n = 6.
13.     Predict the next formula to follow formula (15).

We can also prove formula (13) by using formula (11), the recursive rule for the Pell sequence.

Given Formula (11):   $P_n = 2 \cdot P_{n-1} + 1 \cdot P_{n-2} = P_2 \cdot P_{n-1} + P_1 \cdot P_{n-2}$
Use the rule and replace $P_{n-1}$ with $2 \cdot P_{n-2} + 1 \cdot P_{n-3}$.
Therefore,

$$P_n = 2 \cdot P_{n-1} + 1 \cdot P_{n-2}$$
$$= 2 \cdot [2 \cdot P_{n-2} + 1 \cdot P_{n-3}] + 1 \cdot P_{n-2}$$
$$= 4 \cdot P_{n-2} + 2 \cdot P_{n-3} + 1 \cdot P_{n-2}$$
$$= 5 \cdot P_{n-2} + 2 \cdot P_{n-3}$$

Now replace $P_{n-2}$ with $2 \cdot P_{n-3} + 1 \cdot P_{n-4}$, to get

$$5 \cdot P_{n-2} + 2 \cdot P_{n-3} = 5 \cdot [2 \cdot P_{n-3} + 1 \cdot P_{n-4}] + 2 \cdot P_{n-3}$$
$$= 10 \cdot P_{n-3} + 5 \cdot P_{n-4} + 2 \cdot P_{n-3}$$
$$= 12 \cdot P_{n-3} + 5 \cdot P_{n-4}$$

Therefore,
$$P_n = 12 \cdot P_{n-3} + 5 \cdot P_{n-4} = P_4 \cdot P_{n-3} + P_3 \cdot P_{n-4}, \quad n \geq 4 \quad \blacksquare$$

14.     Prove formula (12) by using formula (11), the recursive rule for the Pell sequence.

**In general, for given n and j, $j \leq n$,**

$$P_n = P_j \cdot P_{n-(j-1)} + P_{j-1} \cdot P_{n-j}, \quad n \geq j \quad \quad (16)$$

15.     Let j = 7 and write out the formula (16).
16.     Verify formula (16) for n = 8 and j = 7.

Consider the **r**th row of the Duplex Triangle (which consists of r + 1 elements), then choose r + 1 consecutive Pell numbers from the Pell sequence. Now form the string dot product of the reverse of the **r**th row and the string of **r + 1** consecutive Pell numbers.  If the last Pell number in the string equals  **p**  and is in position n of the Pell sequence, then the result of the product equals  $P_{r+n}$ .

> **Example:**  Choose row 3  and 3 + 1 = 4  consecutive Pell numbers starting with 5, that is,  (5, 12, 29, 70).  Then **p = 70** and  is, therefore, in position 6.   Thus,
> $P_{3+6} = P_9 = $ **985**.   Reverse the 3$^{rd}$ row of the Duplex Triangle to get  (1, 6, 12, 8).   The result of the string dot product is also 985.
>
> $$(1, 6, 12, 8) \bullet (5, 12, 29, 70) \ = \ 1 \cdot 5 + \ 6 \cdot 12 + \ 12 \cdot 29 \ + \ 8 \cdot 70$$
> $$= \ 5 \ + \ 72 \ + \ 348 \ + \ 560$$
> $$= \ 985$$

---

17.    Form the string of the reverse of the elements in row 2 of the Duplex Triangle and the string (12, 29, 70) of 3 consecutive elements of the Pell sequence.  Find the string dot product of these two strings and show that it is equal to $P_8 = $ **408.**

18.    Choose row 3 of the Duplex Triangle and 3 + 1 = 4  consecutive Pell numbers starting with $P_2 = $ **2.**  Form the string dot product of the reverse of row 3 of the Duplex Triangle and the string of 4 consecutive Pell numbers starting with $P_2 = $ **2.**
Verify that the position of last Pell number in the string is in position n = 5 and that the result of the string dot product is equal to  $P_{3+5} = P_8 = $ **408.**

19.    Form the string consisting of $P_4$ and $P_5$ .  Use row 1 of the Duplex Triangle to form the second string (in reverse) and show that the string dot product of the two strings is equal to $P_6$ .

---

There are other properties of the Pell numbers that fit into patterns that also work for the Fibonacci numbers.  For example, $(F_n)^2 - (F_{n-1}) \cdot (F_{n+1}) = \pm 1$ is true for Fibonacci numbers as is the formula for Pell numbers:

$$(P_n)^2 - (P_{n-1}) \cdot (P_{n+1}) = \pm 1.$$

And the formula $(F_n + 1) \cdot (F_n - 1) = (F_{n-1}) \cdot (F_{n+1})$ is true if n is an odd number greater than 1, while the  corresponding formula for Pell numbers is

$$(P_n + 1) \cdot (P_n - 1) = (P_{n-1}) \cdot (P_{n+1})$$

and is true if n is an odd number greater than 1.

---

20.    Verify the formula  $(P_n)^2 - (P_{n-1}) \cdot (P_{n+1}) = \pm 1$  for n = 3, 4, and 5.

21.    Choose any odd number greater than 1 for n and show that
$$(P_n + 1) \cdot (P_n - 1) = (P_{n-1}) \cdot (P_{n+1}).$$

---

The formulas $(F_n)^2 + (F_{n+1})^2 = F_{2n+1}$ and $F_{2n} = F_n(F_{n-1} + F_{n+1})$ are true for Fibonacci numbers and the corresponding formulas for Pell numbers are also true.

$$(P_n)^2 + (P_{n+1})^2 = P_{2n+1} \quad \text{and} \quad P_{2n} = P_n(P_{n-1} + P_{n+1})$$

---

22.     Verify the formulas $(P_n)^2 + (P_{n+1})^2 = P_{2n+1}$ and $P_{2n}/P_n = P_{n-1} + P_{n+1}$ for n = 1, 2, and 4.

23.     For n = 3, show that $P_{2n+1}$ is a perfect square equal to $(P_n)^2 + (P_{n+1})^2$.

---

Not every formula that is true for Fibonacci numbers has an analog with the Pell numbers. For instance, the formula for the sum of n Fibonacci numbers,

$$1 + 1 + 2 + 3 + 5 + \ldots + F_n = F_{n+2} - 1 \qquad (17)$$

does not work if we substitute Pell numbers for Fibonacci numbers. There is a formula for the sum of n Pell numbers and it is discussed in the next section. Of course, there are formulas for the Pell numbers that have no counterpart with the Fibonacci numbers. For instance, the expression

$$2 \cdot (P_n)^2 \pm 1 \text{ is a square.}$$

---

24.     Show that the expression $2 \cdot (P_n)^2 \pm 1$ is a square for n = 3, 4, and 5.

---

The product of two consecutive Pell numbers is equal to 2 times the sum of the squares of the Pell numbers from $P_1$ up to the smaller of the two consecutive Pell numbers. That is,

$$P_n \cdot P_{n+1} = 2[(P_1)^2 + (P_2)^2 + (P_3)^2 + \ldots + (P_n)^2]$$

---

25.     Verify the formula above for n = 3, 4, and 5.

---

There is a similar formula for the Fibonacci numbers, without the multiplication by 2.

$$F_n \cdot F_{n+1} = (F_1)^2 + (F_2)^2 + (F_3)^2 + \ldots + (F_n)^2$$

---

26.     Verify the formula above for n = 3, 4, and 5.

---

The string dot product of the string of Fibonacci numbers from the first up to the nth and the string of Pell numbers from the nth down to the first is equal to the difference between the (n + 1)th Pell number and the (n + 1)th Fibonacci number. That is,

$$(1, 1, 2, 3, 5, \ldots, F_n) \cdot (P_n, P_{n-1}, P_{n-2}, \ldots, 12, 5, 2, 1)$$

$$= 1 \cdot P_n + 1 \cdot P_{n-1} + 2 \cdot P_{n-2} + 3 \cdot P_{n-3} + 5 \cdot P_{n-4} +$$
$$\ldots + F_{n-3} \cdot 12 + F_{n-2} \cdot 5 + F_{n-1} \cdot 2 + F_n \cdot 1$$

$$= P_{n+1} - F_{n+1} \tag{18}$$

---

27.    Verify the formula $(1, 1, 2, 3, 5) \cdot (29, 12, 5, 2, 1) = P_6 - F_6$

28.    Verify the formula (18) for $n = 3$ and $n = 4$.

---

There are explicit formulas for the Pell numbers (as well as the Fibonacci numbers). We will develop an explicit formula for the Pell numbers in terms of the Duplex Triangle elements from where they came, the Pell diagonals, shown at the beginning of this chapter. (The same procedure can be used for the Fibonacci numbers by referring to Pascal's Triangle.)   Recall, on the $5^{th}$ Pell diagonal of the Duplex Triangle we have $P_5 = 16 + 12 + 1 = 29$. See the figure on the first page of this chapter. The elements 16, 12, and 1 are equal to $N(4, 0)$, $N(3,1)$ and $N(2, 2)$, respectively. Therefore, $P_5 = N(4, 0) + N(3, 1) + N(2, 2)$. Using the formula $N(d, k) = C(d, k) \cdot 2^{d-k}$, we have

$$P_5 = N(4, 0) + N(3, 1) + N(2, 2)$$

$$= C(4, 0) \cdot 2^{4-0} + C(3, 1) \cdot 2^{3-1} + C(2, 2) \cdot 2^{2-2}$$

$$= C(4, 0) \cdot 2^4 + C(3, 1) \cdot 2^2 + C(2, 2) \cdot 2^0$$

The last formula expresses $P_5$ in terms of the binomial coefficients of Pascal's Triangle and powers of 2.

$$C(4, 0) \cdot 2^4 + C(3, 1) \cdot 2^2 + C(2, 2) \cdot 2^0 = 1 \cdot 16 + 3 \cdot 4 + 1 \cdot 1 = 16 + 12 + 1 = 29$$

The steps to convert $P_n$ to Duplex elements in general are as follows:
1) If n is even divide by 2 to determine the number of terms needed.
   If n is odd, add 1, then divide by 2.
2) Sum the terms $N(n - 1, 0) + N(n - 2, 1) + N(n - 3, 2) + \ldots + N(n - j, j - 1)$
   until the number of terms equals the value determined in step 1.  That happens if $n - j$ equals $j - 1$ or if $n - j$ is one greater than $j - 1$.  In general,

$$P_n = N(n - 1, 0) + N(n - 2, 1) + N(n - 3, 2) + \ldots + N(n - j, j - 1)$$

---

29.    Write the expression for $P_6$ (a) in terms of $N(d, k)$ and
       (b) in terms of $C(d, k)$ and the powers of 2.

30.    Write the expression for $P_7$ in terms of $N(d, k)$.

---

## 4.2 The Pythagorean Connection

Pythagorean triples can be formed from a pair of arbitrary positive integers ($x$, $y$, $x > y$) using **Euclid's formulas**. The sides of the right triangle, (**a, b, c**) are determined by **Euclid's formulas**:

$$a = x^2 - y^2, \quad b = 2xy, \quad \text{and} \quad c = x^2 + y^2$$

where **a** and **b** are the legs of the right triangle and **c** is the hypotenuse, so that

$$c^2 = a^2 + b^2.$$

This can be demonstrated algebraically by squaring the quantities shown in Euclid's formulas above. The triple (**a, b, c**) is called a <u>Pythagorean triple</u>.

If we choose consecutive Fibonacci numbers for $x$ and $y$, then **c**, the hypotenuse, will also be a Fibonacci number. Examine the table below:

| $x$ | $y$ | $a = x^2 - y^2$ | $b = 2xy$ | $c = x^2 + y^2$ |
|---|---|---|---|---|
| $F_3 = 2$ | 1 | 3 | 4 | $5 = F_5$ |
| $F_4 = 3$ | 2 | 5 | 12 | $13 = F_7$ |
| $F_5 = 5$ | 3 | 16 | 30 | $34 = F_9$ |
| $F_6 = 8$ | 5 | 39 | 80 | $89 = F_{11}$ |
| $F_7 = 13$ | 8 | 105 | 208 | $233 = F_{13}$ |
| $\cdots$ | $\cdots$ | | | |

Notice that the subscripts of the Fibonacci numbers in the last column are odd numbers and are equal to the sum of the subscripts of Fibonacci numbers represented by $x$ and $y$. Also, note that the sum, **c** + **b**, and the difference, **c** − **b**, are each the square of a Fibonacci number. You can prove this by doing the algebra for c + b and c − b using Euclid's formulas for c and b and noting that x and y in these cases are consecutive Fibonacci numbers, so that both x − y and x + y are also Fibonacci numbers.

The <u>perimeters</u> of the triangles given above are equal to **a** + **b** + **c**, for a given row of the table. Using Euclid's formulas you can show that **a + b + c = 2x(x + y)**. Noting that x and y are consecutive Fibonacci numbers, $F_{n+1}$ and $F_n$, we see that x + y is the next Fibonacci number, $F_{n+2}$. That gives us the perimeter of the triangle equal to $2 \cdot F_{n+2} \cdot F_{n+1}$. Recall (Section 4.1) the formula

$$F_n \cdot F_{n+1} = (F_1)^2 + (F_2)^2 + (F_3)^2 + \ldots + (F_n)^2.$$

Therefore, we have, for a given row of the table, the following formula:

$$a + b + c = 2 \cdot F_n \cdot F_{n+1} = 2 \cdot [(F_1)^2 + (F_2)^2 + (F_3)^2 + \ldots + (F_n)^2].$$

**Section 4.2**

1. Extend the table above by forming the next line in the sequence.
2. In each line of the table involving the Fibonacci numbers, verify that ⟹

(a) the value of **a** is equal to $F_{n+2} \cdot F_{n-1}$, and

(b) the sum, **c + b,** and the difference, **c − b,** are each the square of a Fibonacci number.

3.    For each row of the table involving the Fibonacci numbers, show that the perimeter of the triangle, $\mathbf{a + b + c} = 2 \cdot F_{n+2} \cdot F_{n+1} = F_{2n+3} - (F_n)^2$.

4.    Use Euclid's formulas for **c** and **b** to show that **c + b** and **c − b** are perfect squares, $(x + y)^2$ and $(x - y)^2$.

5.    For each row of the table involving the Fibonacci numbers, show that **a + b + c** for that row is equal to **b** in the next row.

6.    For each row of the table involving the Fibonacci numbers on the previous page, show that the area of the triangle, **(a· b)/2,** for that row is equal to the product of 4 consecutive Fibonacci numbers, that is, Area = $F_{n+2} \cdot F_{n+1} \cdot F_n \cdot F_{n-1}$, and  Area = $(1/2) \cdot (F_n \cdot F_{n-1}) \cdot$ Perimeter.

Now if we choose consecutive Pell numbers for **x** and **y,** in Euclid's formulas then **c,** the hypotenuse, will also be a Pell number.   The triple **(a, b, c)** formed this way will be a "primitive" Pythagorean Triple, meaning that the triple has no common divisor.   Examine the table below:

| x | y | $a = x^2 - y^2$ | $b = 2xy$ | $c = x^2 + y^2$ |
|---|---|---|---|---|
| $P_2 = 2$ | 1 | 3 | 4 | $5 = P_3$ |
| $P_3 = 5$ | 2 | 21 | 20 | $29 = P_5$ |
| $P_4 = 12$ | 5 | 119 | 120 | $169 = P_7$ |
| $P_5 = 29$ | 12 | 697 | 696 | $985 = P_9$ |
| . . . | . . . | | | |

Notice that the subscripts of the Pell numbers in the last column are odd numbers and are equal to the sum of the subscripts of Pell numbers represented by **x** and **y.**  The right triangles whose sides are **a, b** and **c** are "near isosceles", due to the fact that **a** and **b** differ by just 1 unit.  Also, note that the sum, **c + b,** and the difference, **c − b,** are squares of the numbers in the sequence  1, 3, 7, 17, 41, . . ., $q_n$, . . ., where $q_n = 2 \cdot q_{n-1} + q_{n-2}$. This sequence is called "one-half the companion Pell sequence" and plays a role in the discussion of the <u>Companion Pell Sequence</u> to be discussed in the next section, Section 4.3.

7.    Extend the table above, involving the Pell numbers, by forming the next line in the sequence.

8.    Verify that $P_{11}$ is equal to the value of **c** in your extension of the table involving the Pell numbers.

9.    Verify that the sum, **c + b,** and the difference, **c − b,** in each line of the ⟹

table involving the Pell numbers, is the square of a number in the sequence called the "one-half companion Pell sequence", that is, **1, 3, 7, 17, 41, . . . , q$_n$, . . . .**

10. For each row of the table involving the Pell numbers, show that the perimeter of the triangle, **a + b + c,** for that row is equal to a Pell number.

11. (a) Show that the first few elements of the sequence called the "one-half companion Pell sequence", that is, **1, 3, 7, 17, 41, . . ., q$_n$, . . . ,** are equal to sums of two consecutive Pell numbers.
    (b) Show that the value of **a** is the product of two consecutive elements of the sequence called the "one-half companion Pell sequence",

12. For each row k, of the table involving the Pell numbers on the previous page, show that the area of the triangle, **(a· b)/2,** for that row, is equal to the product of 2 consecutive Pell numbers times the product of 2 consecutive **q$_n$** numbers, that is, Area = **P$_{n+1}$ · P$_n$ · q$_{n+1}$ · q$_n$.**

13. For each row k, of the table involving the Pell numbers on the previous page, show that the area of the triangle, **(a· b)/2,** for that row, is Area = **(1/4)· P$_{2n}$ · P$_{2n+2}$**, and Area = **(1/4)·P$_{2n}$·Perimeter.**

From exercises 10 and 13 above, you can deduce that the perimeter of the triangle,

$$\textbf{a + b + c = P}_{2n+2}.$$

Another property of Pell numbers is that the ratio of consecutive Pell numbers form a sequence that approaches what is called the "**silver ratio**", which is equal to $\sqrt{2}+1$. In this respect, the Pell sequence is similar to the Fibonacci sequence, in that the ratio of consecutive Fibonacci numbers form a sequence that approaches the "golden ratio", one of the classical number constants of mathematics.

14. Find the decimal value of each of the following ratios of consecutive Pell numbers: 2/1, 5/2, 12/5, 29/12, 70/29, 169/70

15. Form the next two ratios for the sequence in exercise 5.

16. Compare the decimal values found in exercises 5 and 6 to the decimal value of $\sqrt{2}+1$. Which ratio came closest to $\sqrt{2}+1$?

17. Find the decimal value of **F$_9$/F$_8$** to approximate the value of the golden ratio.

## 4.3 The Companion Pell and Fibonacci Number Sequences

In this and the next section we examine two additional sequences related to the Pell sequence and the Fibonacci sequence. The **Companion Pell numbers**, like the Pell numbers, is an infinite sequence of integers that conforms to the same recursive formula as the Pell sequence, but uses different starting values for the first two elements in the sequence. The companion sequence is

$$2, 2, 6, 14, 34, 82, 198, 478, 1154, \ldots, Q_n, \ldots$$

where **n = 0, 1, 2, 3, . . .**, and $Q_n$ represents the nth number in the sequence and follows the recursive rule

$$Q_n = 2 \cdot Q_{n-1} + Q_{n-2}$$

for $n = 2, 3, \ldots$ and $Q_0 = 2$ and $Q_1 = 2$.

For example, $Q_5 = 2 \cdot Q_4 + Q_3$, or equivalently, $82 = 2 \cdot 34 + 14$.

Many of the formulas valid for the Pell numbers are also valid for the Companion Pell numbers. For instance, the Duplex extensions of the recursive rule:

$$
\begin{aligned}
Q_n &= 1 \cdot Q_n \\
&= 2 \cdot Q_{n-1} + 1 \cdot Q_{n-2} \\
&= 4 \cdot Q_{n-2} + 4 \cdot Q_{n-3} + 1 \cdot Q_{n-4} \\
&= 8 \cdot Q_{n-3} + 12 \cdot Q_{n-4} + 6 \cdot Q_{n-5} + 1 \cdot Q_{n-6} \\
&= 16 \cdot Q_{n-4} + 32 \cdot Q_{n-5} + 24 \cdot Q_{n-6} + 8 \cdot Q_{n-7} + 1 \cdot Q_{n-8} \\
&= 32 \cdot Q_{n-5} + 80 \cdot Q_{n-6} + 80 \cdot Q_{n-7} + 40 \cdot Q_{n-8} + 10 \cdot Q_{n-9} + 1 \cdot Q_{n-10}
\end{aligned}
$$

etc.

The formulas above are recursive and represent the <u>string dot products</u> of the rows of the Duplex Triangle with previous consecutive elements of the Companion Pell sequence. They are analogous to formulas (6) through (10) for the Pell sequence.

The significance of the Pell numbers is that they form the denominators of a sequence of rational numbers that approximate the square root of 2. The numerators of this sequence are equal to one-half of the Companion Pell numbers, that is,

$$1, 3, 7, 17, 41, \ldots, q_n, \ldots, \text{ where } q_n = 2 \cdot q_{n-1} + q_{n-2}.$$

The sequence of approximations begins $1/1$, $3/2$, $7/5$, $17/12$, $41/29$, . . . .

### Section 4.3

1.    Verify the formula $Q_n = 4 \cdot Q_{n-2} + 4 \cdot Q_{n-3} + 1 \cdot Q_{n-4}$ for $n = 5$.
2.    Compare the square root of 2 to the five fractions in the discussion above. Which one is closest in value to the square root of 2?
3.    Find the next rational number (following $41/29$) in the sequence of approximations for the square root of 2. Is this ratio closer to the square root of 2 than the one determined in exercise 2?
4.    Verify the formula $q_n = P_n + P_{n-1}$, for $n = 3, 4$, and 5.

There are formulas that relate the Pell numbers to the Companion Pell numbers. For instance, the sum and difference between two consecutive Companion Pell numbers are each equal to 4 times a Pell number. The formulas are

$$4 \cdot P_n = Q_n + Q_{n-1} \quad \text{and} \quad 4 \cdot P_n = Q_{n+1} - Q_n . \qquad \text{(19) and (19a)}$$

Other formulas are

$$(Q_{n+1})^2 + (Q_n)^2 = 8 \cdot P_{2n+1} \quad \text{and} \quad (Q_{n+1})^2 - (Q_n)^2 = 16 \cdot P_{n+1} \cdot P_n , \quad \text{(20)}$$

and

$$Q_n = P_{n+1} + P_{n-1,}$$

and

$$P_n \cdot Q_n = P_{2n} .$$

---

5.  Verify the six formulas above for $n = 3, 4,$ and 5.
6.  Use formulas (19) and (19a) to show that $8 \cdot P_n = Q_{n-1} + Q_{n+1}$.
7.  Use formula (20) and the formula of Section 4.1, Exercise 25 to show that
    $$(Q_{n+1})^2 - (Q_n)^2 = 32[(P_1)^2 + (P_2)^2 + (P_3)^2 + \ldots + (P_n)^2].$$

---

The <u>string dot product</u> of the string of Pell numbers from the first up to the nth and the string of Companion Pell numbers from the first up to the nth is equal to one less than the (2n + 1)th Pell number divided by 2.  For example, form the strings of the first three Pell numbers and the first three Companion Pell numbers and compute their string dot product. The result is the $7^{\text{th}}$ Pell number ($2n + 1 = 2 \cdot 3 + 1 = 7$) minus 1 divided by 2.

$$(1, 2, 5) \bullet (2, 6, 14) = 2 + 12 + 70 = 84$$

and

$$(P_7 - 1)/2 = (169 - 1)/2 = 168/2 = 84$$

In general,

$$(1, 2, 5, \ldots, P_n) \bullet (2, 6, 14, \ldots, Q_n) = (P_{2n+1} - 1)/2 .$$

---

8.  Verify the general formula above for $n = 4$ and 5.

---

## The Pell Companion–Pythagorean Triples

Recall the sides of a right triangle, **(a, b, c)**, are determined by **Euclid's formulas:**

$$a = x^2 - y^2, \quad b = 2xy, \quad \text{and} \quad c = x^2 + y^2$$

where $x > y$ and **a** and **b** are the legs of the right triangle and **c** is the hypotenuse, so that **a, b,** and **c** satisfy the Pythagorean formula:

$$c^2 = a^2 + b^2.$$

Now if we choose consecutive Companion Pell numbers for x and y, then **c**, the hypotenuse, will be a function of the Pell numbers.  Examine the table on the next page:

| x | y | $a = x^2 - y^2$ | $b = 2xy$ | $c = x^2 + y^2$ |
|---|---|---|---|---|
| $Q_2 = 6$ | 2 | 32 | 24 | $40 = 8 \cdot P_3$ |
| $Q_3 = 14$ | 6 | 160 | 168 | $232 = 8 \cdot P_5$ |
| $Q_4 = 34$ | 14 | 960 | 952 | $1352 = 8 \cdot P_7$ |
| $Q_5 = 82$ | 34 | 5568 | 5576 | $7880 = 8 \cdot P_9$ |
| . . . | . . . | | | |

Notice that the value of **c** is a multiple of a Pell number, that is, $8 \cdot P_k$, also the subscripts of the Pell numbers in the last column are odd numbers and are equal to the sum of the subscripts of the Companion Pell numbers represented by **x** and **y**.    Notice that the difference, **b − a**, is **± 8**.  The value of **a** is also a function of Pell numbers,

$$a = 16 \cdot P_{n+1} \cdot P_n .$$

---

9.   Extend the table above, involving the Companion Pell numbers, by forming the next line in the sequence.

10.   Verify that $8 \cdot P_{11}$ is equal to the value of **c** in your extension of the table involving the Companion Pell numbers and that $16 \cdot P_6 \cdot P_5$ is equal to the value of **a**.

11.   For each row of the table involving the Companion Pell numbers, show that the perimeter of the triangle, **a + b + c,** for that row is equal to $8 \cdot P_{2n+2}$.

12.   For each row k,  of the table involving the Companion Pell numbers, show that the area of the triangle, **(a· b)/2,**  for that row, is
Area = $2 \cdot P_{2n} \cdot$ Perimeter.

---

From exercises 11 and 12 above, you can deduce that the area of the triangle,

$$(a \cdot b)/2 = 16 \cdot P_{2n} P_{2n+2}.$$

## 4.4  The Lucas Numbers

The Companion Pell numbers are sometimes called the **Pell-Lucas** numbers.  This is because there is a companion sequence to the Fibonacci sequence called the sequence of **Lucas numbers** (Edouard Lucas, 1842 - 1891), and it, similar to the Companion Pell sequence, uses the same recursive formula as the Fibonacci sequence, but with different starting values. The sequence  of Lucas numbers is

$$1, 3, 4, 7, 11, 18, 29, 47, 76, 123, \ldots, L_n, \ldots$$

where **n = 1, 2, 3,** $\ldots$, and $L_n$ represents the nth number in the sequence and follows the recursive rule

$$L_n = L_{n-1} + L_{n-2}$$

for n = 3, 4, . . . and $L_1 = 1$ and $L_2 = 3$.

For example, $L_5 = L_4 + L_3$, or equivalently, 11 = 7 + 4.

Many of the formulas valid for the Fibonacci numbers are also valid for the Companion Fibonacci numbers, or Lucas numbers. For instance, the Pascal extensions of the recursive rule:

$$L_n = 1 \cdot L_n$$
$$= 1 \cdot L_{n-1} + 1 \cdot L_{n-2}$$
$$= 1 \cdot L_{n-2} + 2 \cdot L_{n-3} + 1 \cdot L_{n-4}$$
$$= 1 \cdot L_{n-3} + 3 \cdot L_{n-4} + 3 \cdot L_{n-5} + 1 \cdot L_{n-6}$$
$$= 1 \cdot L_{n-4} + 4 \cdot L_{n-5} + 6 \cdot L_{n-6} + 4 \cdot L_{n-7} + 1 \cdot L_{n-8}$$
$$= 1 \cdot L_{n-5} + 5 \cdot L_{n-6} + 10 \cdot L_{n-7} + 10 \cdot L_{n-8} + 5 \cdot L_{n-9} + 1 \cdot L_{n-10}$$

**etc.**

The preceding formulas are recursive and represent the <u>string dot products</u> of the rows of Pascal's Triangle with previous consecutive elements of the Lucas sequence. They are analogous to formulas (1) through (5) for the Fibonacci sequence.

### Section 4.4

1.      Verify the formula $L_n = 1 \cdot L_{n-3} + 3 \cdot L_{n-4} + 3 \cdot L_{n-5} + 1 \cdot L_{n-6}$ for n = 8.

There are formulas that relate the Lucas numbers to the Fibonacci numbers. For instance,

$$L_n = F_{n+1} + F_{n-1},$$
$$L_n = F_n + 2 \cdot F_{n-1},$$
$$(L_{n+1})^2 - (L_n)^2 = (L_{n-1}) \cdot (L_{n+2}),$$
$$(L_n) \cdot (L_{n+1}) = L_{2n+1} + (-1)^{n+1} = F_{2n+2} + F_{2n} + (-1)^n,$$

and

$$(L_{n+1})^2 + (L_n)^2 = L_{2n+2} + L_{2n} = 5 \cdot F_{2n+1}.$$

2.      Verify the five formulas above for n = 3, 4, and 5.

There is an interesting connection between the Duplex Triangle and the Lucas numbers. Form the string consisting of the elements of the $3^{rd}$ row of the Duplex Triangle in reverse order and the string of the Lucas numbers from the first to the fourth (there are 4 elements in the 3$^{rd}$ row of the Duplex Triangle). Then find the <u>string dot product</u> to get the 10$^{th}$ Lucas number.

$$L_{10} = (1, 6, 12, 8) \cdot (1, 3, 4, 7)$$
$$= 1 + 18 + 48 + 56$$
$$= 123$$

We find this to be the Lucas number in the 10$^{th}$ position of the sequence by multiplying the row number used in the first string by 3 and adding 1, that is, $3 \cdot 3 + 1 = 10$.

*In general, choose row r of the Duplex Triangle and reverse it for the first string. Then choose a string of (r + 1) consecutive Lucas numbers, noting the index of the smallest or first Lucas number in the string, say k. Then the string dot product of these two strings equals the (3 · r + k)th Lucas number.*

**Example:** Choose row 2 of the Duplex Triangle and reverse it. Then choose $2 + 1 = 3$ consecutive Lucas numbers starting with 7, that is, $(7, 11, 18)$. The string dot product is

$$(1, 4, 4) \cdot (7, 11, 18) = 7 + 44 + 72 = 123 = L_{10}$$

The subscript 10 is found by multiplying 3 times the row number chosen and adding the index position of the first Lucas number in the second string which is 4, since $7 = L_4$. Thus, $3 \cdot 2 + 4 = 10$.

---

3.    Find the string dot product of the string consisting of the reverse of the 3$^{rd}$ row of the Duplex Triangle and 4 consecutive Lucas numbers starting with $3 = L_2$. Verify that the result is $L_{11}$.

4.    Choose your own row of the Duplex Triangle and your own string of consecutive Lucas numbers consistent with the length of the first string corresponding to the row you chose. Find the Lucas number and its index that equals the string dot product of your two strings.

---

## The Lucas–Pythagorean Triples

Recall the sides of a right triangle, **(a, b, c)**, are determined by **Euclid's formulas:**

$$a = x^2 - y^2, \quad b = 2xy, \quad \text{and} \quad c = x^2 + y^2$$

where $x > y$ and **a** and **b** are the legs of the right triangle and **c** is the hypotenuse, so that **a, b,** and **c** satisfy the Pythagorean formula:

$$c^2 = a^2 + b^2.$$

Now if we choose consecutive Lucas numbers for x and y, then **c**, the hypotenuse, will be a function of Lucas numbers or Fibonacci numbers.
Examine the table below:

| x | y | $a = x^2 - y^2$ | $b = 2xy$ | $c = x^2 + y^2$ |
|---|---|---|---|---|
| $L_2 =$   3 | 1 | 8 | 6 | $10 = L_5 - L_1 = L_4 + L_2 = 5 \cdot F_3$ |
| $L_3 =$   4 | 3 | 7 | 24 | $25 = L_7 - L_3 = L_6 + L_4 = 5 \cdot F_5$ |
| $L_4 =$   7 | 4 | 33 | 56 | $65 = L_9 - L_5 = L_8 + L_6 = 5 \cdot F_7$ |
| $L_5 =$   11 | 7 | 72 | 154 | $170 = L_{11} - L_7 = L_{10} + L_8 = 5 \cdot F_9$ |
| $L_6 =$   18 | 11 | 203 | 396 | $445 = L_{13} - L_9 = L_{12} + L_{10} = 5 \cdot F_{11}$ |
| $L_7 =$   29 | 18 | 517 | 1044 | $1165 = L_{15} - L_{11} = L_{14} + L_{12} = 5 \cdot F_{13}$ |
| $\cdots$ | $\cdots$ | | | |

Notice that the value of **c** is a multiple of **5**, that is, $5 \cdot F_k$, a multiple of a Fibonacci number, and the subscript k is equal to the sum of the subscripts of the Lucas numbers numbers represented by x and y. Note that the value of **b** in row k of the table is 2 times the quantity $(L_{2k+1} + (-1)^k)$, that is, 2 times the quantity, a Lucas number plus or minus 1. Also, note that the value of **a** is the product of two Lucas numbers whose subscripts differ by 3, for example, $33 = (L_2)(L_5)$ and $72 = (L_3)(L_6)$ (for row 1 assume $L_0 = 2$).

---

5.   Extend the table of Pythagorean triples above by forming the next line in the sequence.

6.   Verify that $5 \cdot F_{15}$ is equal to the value of **c** in your extension of the table.

7.   Verify that the value of **b**, in each row k of the table of Pythagorean triples above, is equal to $2 \cdot (L_{2k+1} + (-1)^k)$.

8.   Verify that the value of **a**, in each row $k > 1$ of the table of Pythagorean triples above, is equal to the product of two Lucas numbers whose subscripts differ by 3 (for row 1 assume $L_0 = 2$).

9.   For each row of the table of Pythagorean triples above, show that the triangle perimeter, **a + b + c**, for that row is equal to
   (a)   the value of **b** in the next row, or $2 \cdot (L_{2k+3} + (-1)^{k+1})$, and
   (b)   $2 \cdot (F_{2k+4} + F_{2k+2} + (-1)^{k+1})$, and
   (c)   $2 \cdot (L_{2k+3} + (-1)^{k+1})$, and
   (d)   $2 \cdot L_{k+1} \cdot L_{k+2}$ .

10.   For each row k of the table of Pythagorean triples above, show that the area of the triangle, **(a · b)/2**, for that row is equal to
   (a)   the product of 4 consecutive Lucas numbers, starting with $L_{k-1}$   ⟹
      (for row 1 assume $L_0 = 2$), and

**(b)  $(1/2) \cdot (L_k \cdot L_{k-1}) \cdot$Perimeter.**

## 4.5   The Sum of the Pell Numbers

The sum of Pell numbers from the first to the nth,  $1 + 2 + 5 + 12 + 29 + \ldots + P_n$ is equal to one-forth the quantity of $Q_{n+1}$ minus 2.  The formula is

$$1 + 2 + 5 + 12 + 29 + \ldots + P_n = (Q_{n+1} - 2)/4 \qquad (21)$$

Another formula,  somewhat similar to the formula for the sum of the Fibonacci numbers,  is

$$1 + 2 + 5 + 12 + 29 + \ldots + P_n = (P_{n+1} + P_n - 1)/2. \qquad (22)$$

Formula (17) gives the sum of the Fibonacci numbers as  $F_{n+2} - 1$, but by the recursive rule, $F_{n+2}$ is equal to the sum of the two previous Fibonacci numbers.  Therefore, we have the following comparison between the  sum of the Fibonacci numbers and the sum of the Pell numbers:

**sum of Fibonacci numbers $= F_{n+2} - 1 = F_{n+1} + F_n - 1$,**

**and sum of Pell numbers $= (P_{n+1} + P_n - 1)/2$ .**

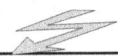

**Section  4.5**

1.      Use the two formulas (21) and (22) to find the sum  **$1 + 2 + 5 + 12 + 29 + 70$.** Verify that each formula yields the same result.

2.      Find the sum  **$1 + 2 + 5 + 12 + 29 + \ldots + P_n$**  for $n = 3, 5$, and 8 using one of the formulas above.

3.      Show that  **$(P_{n+1} + P_n - 1)/2 = (Q_{n+1} - 2)/4$**  using formula (20) and the recursive rule for Pell numbers.

The sum of the Pell numbers from the first up to a multiple of 4 plus 1 is always a perfect square.  For example,  the sum of the first 5 Pell numbers is equal to 7 squared,

$$1 + 2 + 5 + 12 + 29 = 49 = 7^2.$$

Furthermore, 7 is the sum of the second and third Pell numbers.

4.      Show that the sum  **$1 + 2 + 5 + 12 + 29 + 70 + 169 + 408 + 985$**  is a perfect square equal to 41 squared.  Furthermore, show that 41 is the sum of the $4^{th}$ and $5^{th}$ Pell numbers.

In general,

$$1 + 2 + 5 + 12 + 29 + \ldots + P_{4n+1} = (P_{2n+1} + P_{2n})^2 .$$

---

5.    Complete the following partial sums of the $P_n$ numbers.

$$1 = 1$$
$$1 + 2 = 3$$
$$1 + 2 + 5 = 8$$
$$1 + 2 + 5 + 12 = 20$$
$$1 + 2 + 5 + 12 + 29 = 49$$
$$1 + 2 + 5 + 12 + 29 + 70 = \underline{\quad}$$
$$1 + 2 + 5 + 12 + 29 + 70 + 169 = \underline{\quad}$$

---

The sequence of partial sums of the Pell numbers (started in Exercise 5 above) is also a recursive sequence:

$$1, 3, 8, 20, 49, \ldots , T_n , \ldots ,$$

where $\qquad T_n = 2 \cdot T_{n-1} + 1 \cdot T_{n-2} + 1$

and $T_1 = 1$ and $T_2 = 3$ and $n = 3, 4, 5, \ldots$.

---

6.    Verify the formula $T_n = 2 \cdot T_{n-1} + 1 \cdot T_{n-2} + 1$ for $n = 3, 4,$ and 5.
7.    Verify the formula $T_n = (P_{n+1} + P_n - 1)/2$ for $n = 3, 4,$ and 5.
8.    Verify the formula $P_n = T_{n-1} + T_{n-2} + 1$ for $n = 3, 4,$ and 5.

---

The sequence of partial sums of the Fibonacci numbers is also a recursive sequence:

$$1, 2, 4, 7, 12, 20, 33, \ldots , G_n , \ldots ,$$

where $\qquad G_n = G_{n-1} + G_{n-2} + 1$ and $G_1 = 1$ and $G_2 = 2$ and $n = 3, 4, 5, \ldots$.

---

9.    Make table of partial sums of the Fibonacci numbers similar to the table for the Pell numbers in Exercise 5.
10.   Verify the formula $G_n = G_{n-1} + G_{n-2} + 1$ for $n = 6, 7,$ and 8.
11.   Verify the formula $G_n = F_{n+1} + F_n - 1$ for $n = 3, 4,$ and 5.
12.   Given that $G_n = F_{n+1} + F_n - 1$ show that $G_n = F_{n+2} - 1$.
13.   Verify the formula $F_n = G_{n-2} + 1$ for $n = 3, 4,$ and 5.

## 4.6   Related Sequences of the Pell and Fibonacci Numbers

The objectives of this chapter has been to not only show the connection between the Pell numbers and the Duplex Triangle, but also to point out some of the relationships and analogies the Pell numbers have with the Fibonacci numbers and their connection with Pascal's Triangle.   We conclude this chapter with another such relationship and analogy. First consider the Fibonacci numbers and Pascal's Triangle. The sum of the Fibonacci numbers from $F_1$ to $F_n$ as given by formula (17) is equal to $F_{n+2} - 1$. Also, the sum of all the elements in Pascal's Triangle from row 0 to row n – 1  is equal to $2^n - 1$.   Since the Fibonacci numbers, as found on the Fibonacci diagonals in Pascal's Triangle, correspond to elements in the rows of Pascal's Triangle, but not all of the elements in all of the rows, the formula for the Pascal Triangle elements not represented by Fibonacci numbers for a given value of n, is equal to

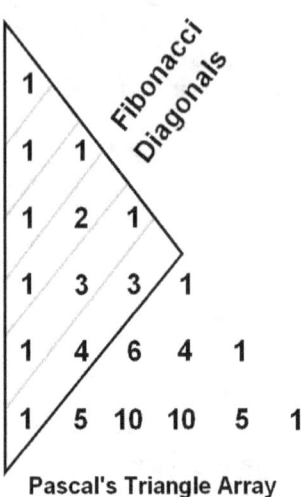

**Pascal's Triangle Array**

$$(2^n - 1) - (F_{n+2} - 1) = 2^n - F_{n+2}.$$

In the figure shown here, we have n = 6 and the sum of the Fibonacci numbers from $F_1$ to $F_6$ equals

$$1 + 1 + 2 + 3 + 5 + 8 = 20,$$

or $\qquad\qquad F_{n+2} - 1 = F_{6+2} - 1 = F_8 - 1 = 20$

The sum of all of the elements in Pascal's Triangle up through the 5th row is equal to $2^6 - 1 = 63$.   The  sum of the remaining elements of Pascal's Triangle, those outside the elements enclosed in the Fibonacci diagonals, is equal to **63 – 20 = 43**.    The expression given above for the sum of those elements is $2^n - F_{n+2}$,  and for n = 6, that is

$$2^n - F_{n+2} = 2^6 - F_{6+2} = 64 - 21 = 43.$$

This is verified by adding all of the elements of Pascal's Triangle shown above that are not on the enclosed Fibonacci diagonals, that is,

$$1 + (6 + 4 + 1) + (5 + 10 + 10 + 5 + 1) = 1 + 11 + 31 = 43.$$

**Section  4.6**

1.   For each of the figures on the next page find  **(a)** the sum of all of the elements shown in Pascal's Triangle, **(b)** the sum of the elements enclosed in triangle, and **(c)** the sum of the elements not enclosed in the triangle. In each case, compare your sums to the values of the expressions
(1) $2^n - 1$,   (2) $F_{n+2} - 1$,  and   (3) $2^n - F_{n+2}$,
where n equals one more than the row number of the last row shown in each figure.

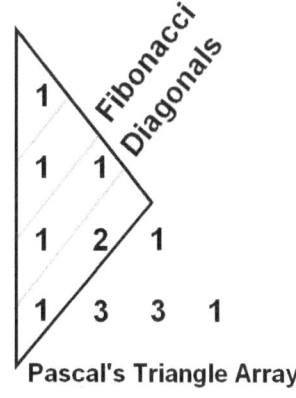

Pascal's Triangle Array

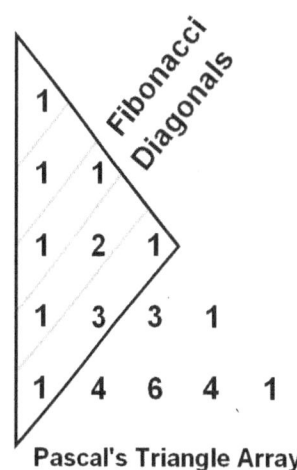

Pascal's Triangle Array

The expression $2^n - F_{n+2}$, given for the sum of those elements of Pascal's Triangle that are **not enclosed** with the Fibonacci diagonals, provides a formula for a sequence of numbers which we will denote by $a(n) = 2^n - F_{n+2}$, for $n = 1, 2, 3, \ldots$.
This produces the sequence

$$0, 1, 3, 8, 19, 43, 94, 201, \ldots, a(n), \ldots,$$

which turns out to have a interesting connection with a coin tossing problem. Suppose you toss a coin n times. How many times will the pattern of heads/tails being generated show at least two consecutive heads? For example, suppose you toss a coin 5 times. Our sequence, yields the answer, that is, the value of $a(n) = 2^n - F_{n+2}$, for $n = 5$.

$$a(5) = 2^5 - F_{5+2} = 32 - F_7 = 32 - 13 = 19.$$

We will list these 19 patterns below. Note: There are only two ways the coin can come up on a single toss, either heads or tails, therefore, there are $2 \times 2 \times 2 \times 2 \times 2 = 2^5 = 32$ possible patterns. It is useful to represent heads – tails with 1's and 0's, and the sequence of tosses with binary number patterns. The 5 outcomes on each trial range from 00000 to 11111 as binary numbers, or, from 5 tails to 5 heads. The list below shows the 19 patterns where there are at least two consecutive heads in the pattern:

00011, 00110, 00111, 01011, 01100, 01101, 01110, 01111,

10011, 10110, 10111, 11011, 11100, 11101, 11110, 11111,

11000, 11001, 11010.

In the list we can observe there are 8 patterns that begin with a 0 (tails). If we replace the beginning 0 with 1, then there are 8 more patterns that now begin with a 1 (heads). Finally, there are 3 additional patterns, all beginning with a 1, that also show at least two consecutive heads in the pattern. It turns out that the numbers 8 and 3 are equal to **a(4)** and $F_4$, respectively, and we can write

$$a(5) = 2 \cdot a(4) + F_4 = 2 \cdot 8 + 3 = 16 + 3 = 19.$$

In general,

$$a(n) = 2 \cdot a(n - 1) + F_{n-1}.$$

This recursive formula fits into the following table:

| n | a(n) | = | $2 \cdot a(n-1) + F_{n-1}$ | = | $2^n - F_{n+2}$ | = a(n) |
|---|---|---|---|---|---|---|
| 1 | 0 | = | $2 \cdot$ 0 + 0 | = | 2 − 2 | = 0 |
| 2 | 1 | = | $2 \cdot$ 0 + 1 | = | 4 − 3 | = 1 |
| 3 | 3 | = | $2 \cdot$ 1 + 1 | = | 8 − 5 | = 3 |
| 4 | 8 | = | $2 \cdot$ 3 + 2 | = | 16 − 8 | = 8 |
| 5 | 19 | = | $2 \cdot$ 8 + 3 | = | 32 − 13 | = 19 |
| 6 | 43 | = | $2 \cdot$ 19 + 5 | = | 64 − 21 | = 43 |
| 7 | 94 | = | ____ + ___ | = | ___ − ___ | = ____ |
| 8 | ____ | = | ____ + ___ | = | ___ − ___ | = ____ |

2.    Complete the table above.

Also, we find that

$$a(n) = 2 \cdot a(n - 1) + F_{n-1} = 2^n - F_{n+2}. \qquad (23)$$

Furthermore, since there are only $2^n$ total patterns for tossing a coin n times, the expression $F_{n+2}$ represents the negation of the requirement of " at least two consecutive heads". Hence, in our example, $F_7 = 13$ is the number of patterns with 5 tosses that **do not** have two consecutive heads.

> In general, $F_{n+2}$ is equal to the number of patterns with n tosses of a coin that <u>do not</u> have two consecutive heads.

3.    Suppose you toss a coin 3 times. How many times will the pattern of heads/tails being generated show at least two consecutive heads? How many patterns will <u>not</u> show two consecutive heads?

4.    Suppose you toss a coin 5 times. How many times will the pattern of heads/tails being generated show at least two consecutive heads? How many patterns will <u>not</u> show two consecutive heads?

5.    Suppose you toss a coin 8 times. How many times will the pattern of heads/tails being generated show at least two consecutive heads? How many patterns will <u>not</u> show two consecutive heads?

We have a relationship with the Duplex Triangle and the Pell numbers similar to the relationship we just explored with Pascal's Triangle and the Fibonacci diagonals. We first look at the sum of the Pell numbers from $P_1$ to $P_n$ as given by formula (22), and that sum is equal to

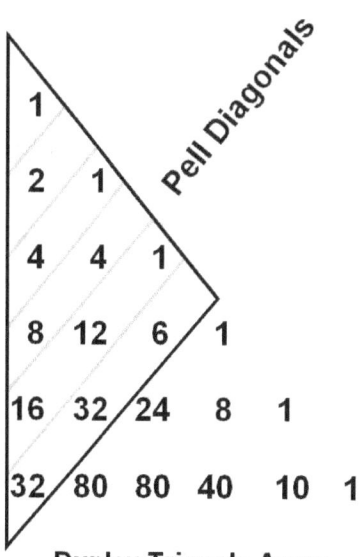

**Duplex Triangle Array**

$$(P_{n+1} + P_n - 1)/2.$$

Also, the sum of all the elements in the Duplex Triangle from row 0 to row $n-1$ is equal to

$$(1/2)(3^n - 1)$$

(refer to Chapter 3). Since the sum of the Pell numbers, as found on the Pell diagonals in the Duplex Triangle correspond to elements enclosed by the triangle shown in the figure, the formula for the sum of the Duplex Triangle elements **not** represented by Pell numbers for a given value of n, is equal to the sum of all the elements of the Duplex Triangle up to row $(n-1)$ minus the sum of the Pell numbers from $P_1$ to $P_n$. That is,

$$(1/2)(3^n - 1) - (P_{n+1} + P_n - 1)/2.$$

In the figure shown above, we have n = 6 and the sum of the Pell numbers from $P_1$ to $P_6$ (those elements inside the drawn triangle) equals

$$1 + 2 + 5 + 12 + 29 + 70 = 119,$$

or, by the formula,

$$(P_{n+1} + P_n - 1)/2 = (P_7 + P_6 - 1)/2 = (70 + 169 - 1)/2 = 119$$

The sum of all of the elements in the Duplex Triangle up through the 5th row is equal to

$$(1/2)(3^n - 1) = (1/2)(3^6 - 1) = (1/2)(729 - 1) = 364.$$

The sum of the remaining elements of the Duplex Triangle, those outside the elements enclosed in the Pell diagonals, is equal to **364 – 119 = 245**.

Therefore, the expression given above for the sum of those elements, that is the formula

$$(1/2)(3^n - 1) - (P_{n+1} + P_n - 1)/2,$$

for n = 6 is

$$(1/2)(3^6 - 1) - (P_{6+1} + P_6 - 1)/2 = 364 - 119 = 245.$$

This is verified by adding all of the elements of the Duplex Triangle shown above that are **not** enclosed by the Pell diagonals:

$$1 + (24 + 8 + 1) + (80 + 80 + 40 + 10 + 1) = 1 + 33 + 211 = 245.$$

6.    For each of the figures below find **(a)** the sum of all of the elements shown in the Duplex Triangle, **(b)** the sum of the elements enclosed in the drawn triangle, and **(c)** the sum of the elements not enclosed in the triangle.  In each case, compare your sums to the values of the expressions

(1)   $(1/2)(3^n - 1)$,          (2)   $(P_{n+1} + P_n - 1)/2$,

and   (3)   $(1/2)(3^n - 1) - (P_{n+1} + P_n - 1)/2$,

where n equals one more than the row number of the last row shown in each figure.

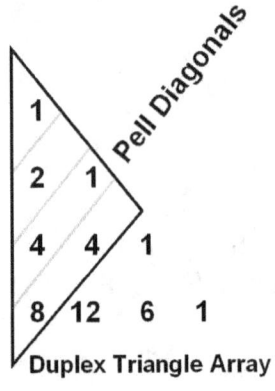

**Duplex Triangle Array**

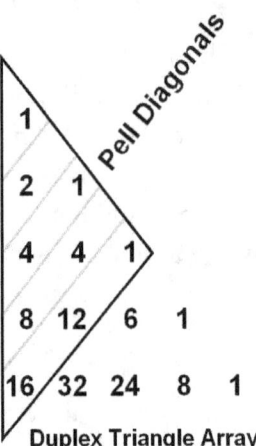

**Duplex Triangle Array**

The expression  $(1/2)(3^n - 1) - (P_{n+1} + P_n - 1)/2$, given for the sum of those elements of the Duplex Triangle that are **not enclosed** with the Pell diagonals, provides a formula for a sequence of numbers which we will denote by

$$b(n) = (1/2)(3^n - 1) - (P_{n+1} + P_n - 1)/2, \text{ for } n = 1, 2, 3, \ldots.$$

This produces the sequence

0, 1, 5, 20, 72, 245, 805, 2584, . . ., b(n), . . . .

Also, notice that $b(4) = 20 = 3 \cdot b(3) + 5$ and $b(5) = 72 = 3 \cdot b(4) + 12$.  The 5 and 12 that were added in those two formulas are the Pell numbers $P_3$ and $P_4$.  Noting this we can write

$$b(4) = 3 \cdot b(3) + P_3 \text{ and } b(5) = 3 \cdot b(4) + P_4.$$

7.    Complete the table on the next page.  Compare the final values of columns 3 and 4 with the values in column 2.  Make a conjecture.

| n | b(n) | $(1/2)(3^n - 1) - (P_{n+1} + P_n - 1)/2$ | $3 \cdot b(n-1) + P_{n-1}$ |
|---|---|---|---|
| 1 | 0 | $(1/2)(3^1 - 1) - (P_2 + P_1 - 1)/2 = \quad 0$ | $3 \cdot b(0) + P_0 \quad = \quad 0$ |
| 2 | 1 | $(1/2)(3^2 - 1) - (P_3 + P_2 - 1)/2 = \quad 1$ | $3 \cdot b(1) + P_1 \quad = \quad 1$ |
| 3 | 5 | $(1/2)(3^3 - 1) - (P_4 + P_3 - 1)/2 = \quad 5$ | $3 \cdot b(2) + P_2 \quad = \quad 5$ |
| 4 | 20 | $(1/2)(3^4 - 1) - (P_5 + P_4 - 1)/2 = \quad 20$ | $3 \cdot b(3) + P_3 \quad = \quad 20$ |
| 5 | 72 | $(1/2)(3^5 - 1) - (P_6 + P_5 - 1)/2 = \underline{\quad}$ | $3 \cdot b(4) + P_4 \quad = \underline{\quad}$ |
| 6 | 245 | $(1/2)(3^6 - 1) - (P_7 + P_6 - 1)/2 = \underline{\quad}$ | $3 \cdot b(5) + P_5 \quad = \underline{\quad}$ |
| 7 | 805 | $\underline{\qquad\qquad\qquad} = \underline{\quad}$ | $\underline{\qquad\qquad} = \underline{\quad}$ |
| 8 | ___ | $\underline{\qquad\qquad\qquad} = \underline{\quad}$ | $\underline{\qquad\qquad} = \underline{\quad}$ |

In general,

$$b(n) = (1/2)(3^n - 1) - (P_{n+1} + P_n - 1)/2 = 3 \cdot b(n-1) + P_{n-1}. \quad (24)$$

Another formula for the general term of the sequence **b(n)** is

$$b(n) = b(n-1) + 3^{n-1} - P_n. \quad (25)$$

---

8.　　Evaluate formula (25) for n = 3, 4, and 5. Verify the values found with those in the table above.

---

It is interesting to note the resemblance of formula (23) for **a(n)**, the sequence associated with Pascal's Triangle and the Fibonacci diagonals and formula (24) for **b(n)**, the sequence associated with the Duplex Triangle and the Pell diagonals.

$$a(n) = 2 \cdot a(n-1) + F_{n-1}$$

$$b(n) = 3 \cdot b(n-1) + P_{n-1}.$$

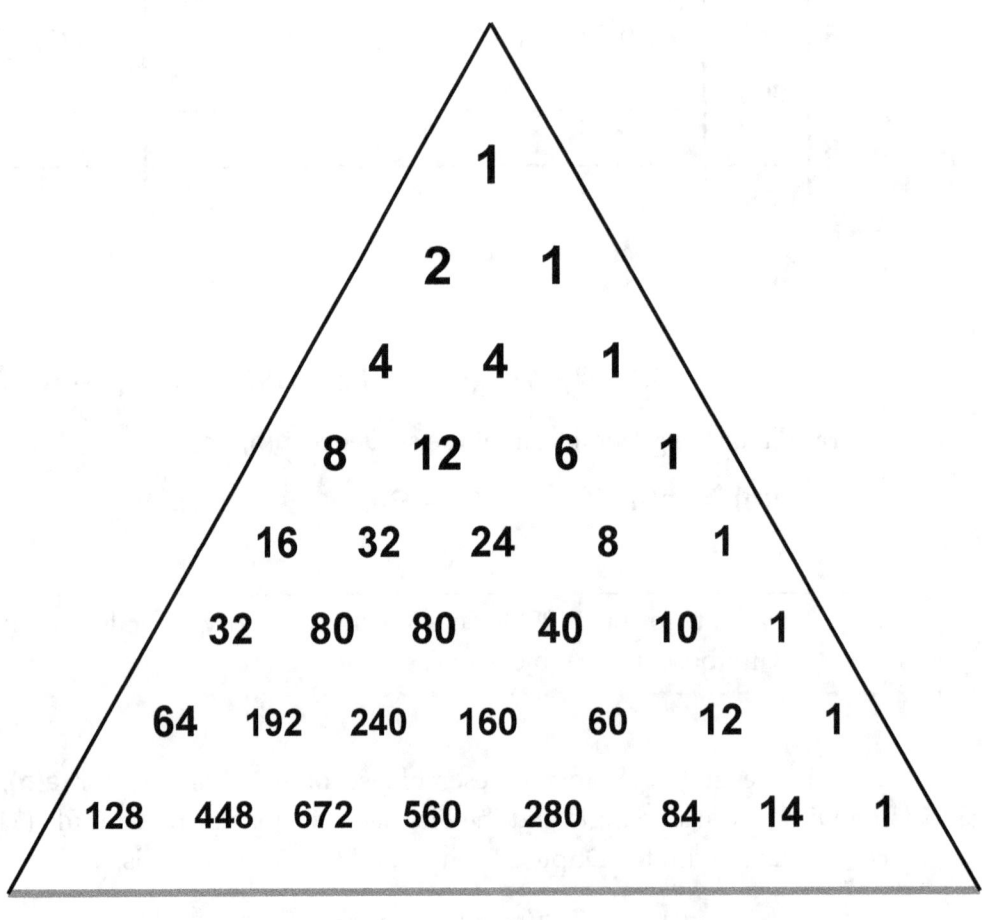

# 5

# Anti-Diagonal Patterns Within the Duplex Triangle
## - The Jacobsthal Numbers

## 5.1 Jacobsthal Numbers

The infinite sequence of integers

$$0, 1, 1, 3, 5, 11, 21, 43, 85, 171, 341, \ldots, J_n, \ldots$$

where **n = 0, 1, 2, 3, . . .,** and $J_n$ represents the nth number in the sequence, is named the Jacobsthal sequence, or the elements in the sequence are called Jacobsthal Numbers (Ernst Jacobsthal, German mathematician, 1882-1965). The significance of the Jacobsthal numbers as they relate to the Duplex Triangle, is that the sums of the elements on the anti-diagonals of the Duplex Triangle are the Jacobsthal numbers. The anti-diagonals are different from the diagonals (Pell diagonals) of the Duplex Triangle, due to the asymmetry of the entries on the rows of the Triangle, unlike the symmetry seen in Pascal's Triangle. See the figure below.

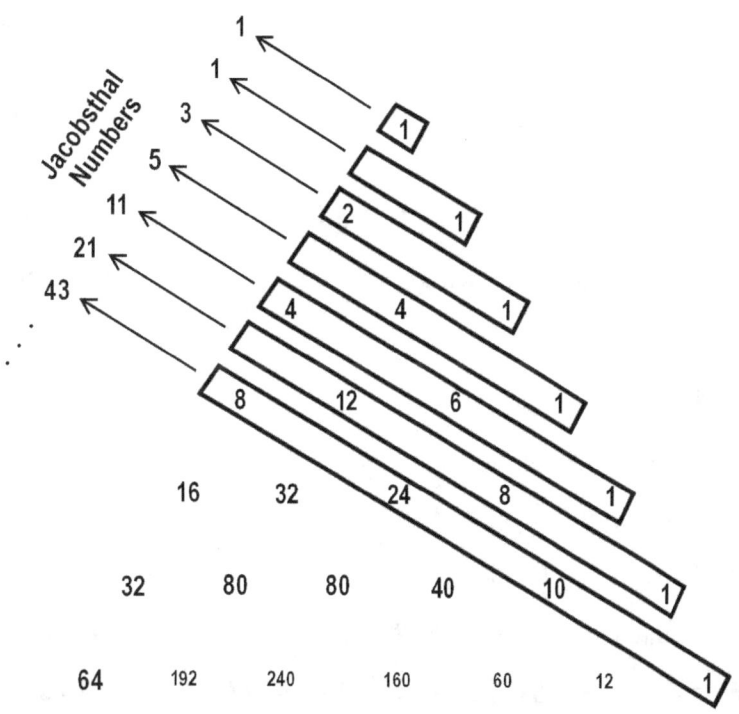

**Jacobsthal Diagonals in the Duplex Triangle**

On the $5^{th}$ anti-diagonal , for example, we have $J_5 = 1 + 6 + 4 = 11$.

Like the Fibonacci and Pell sequences, a formula for the Jacobsthal sequence is a recursive formula where each element is based on the previous two elements in the sequence. The Jacobsthal Rule:

$$J_n = 1 \cdot J_{n-1} + 2 \cdot J_{n-2}$$

where $J_0 = 0$ and $J_1 = 1$ and $n = 2, 3, 4, 5, \ldots$.

For example, $J_5 = 1 \cdot J_4 + 2 \cdot J_3$, or equivalently, $11 = 1 \cdot 5 + 2 \cdot 3$.

The Jacobsthal sequence, like the Fibonacci and Pell sequences, belong to a larger class of integer sequences called Lucas Sequences. They each have recursive rules that depend on the previous two elements in the sequence.

| | | |
|---|---|---|
| **Fibonacci:** | $F_n = 1 \cdot F_{n-1} + 1 \cdot F_{n-2},$ | $F_0 = 0$ and $F_1 = 1$ |
| **Pell:** | $P_n = 2 \cdot P_{n-1} + 1 \cdot P_{n-2},$ | $P_0 = 0$ and $P_1 = 1$ |
| **Jacobsthal:** | $J_n = 1 \cdot J_{n-1} + 2 \cdot J_{n-2},$ | $J_0 = 0$ and $J_1 = 1$ |

Each sequence has a "companion sequence", which is a sequence with the same recursive rule as its friend, but with different initial starting values. The companion Jacobsthal sequence, sometimes called the Jacobsthal-Lucas sequence, has the starting values 2 and 1, respectively, for $n = 0$ and $n = 1$. We will explore this sequence further on in the Chapter.

## 5.2 Duplex Triangle Extensions of the Recursive Rule

The recursive rule for the Jacobsthal numbers is formed from the <u>string dot product</u> of the <u>reverse</u> of row 1 of the Duplex Triangle and two previous consecutive elements of the Jacobsthal sequence, giving us

$$J_n = (1, 2) \bullet (J_{n-1}, J_{n-2}) = 1 \cdot J_{n-1} + 2 \cdot J_{n-2}$$

In general, we have the following formulas, which are the Duplex extensions of the recursive Jacobsthal Rule.

$$
\begin{aligned}
J_n &= 1 \cdot J_n \\
&= 1 \cdot J_{n-1} + 2 \cdot J_{n-2} & (1) \\
&= 1 \cdot J_{n-2} + 4 \cdot J_{n-3} + 4 \cdot J_{n-4} & (2) \\
&= 1 \cdot J_{n-3} + 6 \cdot J_{n-4} + 12 \cdot J_{n-5} + 8 \cdot J_{n-6} & (3) \\
&= 1 \cdot J_{n-4} + 8 \cdot J_{n-5} + 24 \cdot J_{n-6} + 32 \cdot J_{n-7} + 16 \cdot J_{n-8} & (4) \\
&= 1 \cdot J_{n-5} + 10 \cdot J_{n-6} + 40 \cdot J_{n-7} + 80 \cdot J_{n-8} + 80 \cdot J_{n-9} + 32 \cdot J_{n-10} & (5)
\end{aligned}
$$

etc.

These formulas are recursive and represent the <u>string dot products</u> of the <u>reverse</u> of the rows of the Duplex Triangle with previous consecutive elements of the Jacobsthal sequence. For instance, if we use row 2 of the Duplex Triangle in the product, then the consecutive sequence of Jacobsthal numbers begins with $J_{n-2}$ and works backward for two more Jacobsthal numbers, a total of three consecutive Jacobsthal numbers. This assumes that n is at least equal to 4.

$$J_n = (1, 4, 4) \bullet (J_{n-2}, J_{n-3}, J_{n-4}) = 1 \cdot J_{n-2} + 4 \cdot J_{n-3} + 4 \cdot J_{n-4}$$

**Example:** Verify formula (2) for n = 6.
**Solution:** $J_6 = 21$ and

$$1 \cdot J_{n-2} + 4 \cdot J_{n-3} + 4 \cdot J_{n-4} = 1 \cdot J_4 + 4 \cdot J_3 + 4 \cdot J_2$$
$$= 1 \cdot 5 + 4 \cdot 3 + 4 \cdot 1$$
$$= 5 + 12 + 4$$
$$= 21$$

**Note**: If both strings are reversed, the same result is obtained.
$$(4, 4, 1) \bullet (1, 3, 5) = 4 + 12 + 5 = 21.$$

We can also prove formula (2) by using formula (1), the given recursive rule for the Jacobsthal sequence.

Given Formula (1): $J_n = 1 \cdot J_{n-1} + 2 \cdot J_{n-2}$
Use the recursive rule and replace $J_{n-1}$ and $J_{n-2}$ with
$1 \cdot J_{n-2} + 2 \cdot J_{n-3}$ and $1 \cdot J_{n-3} + 2 \cdot J_{n-4}$, respectively.

Therefore,

$$J_n = 1 \cdot J_{n-1} + 2 \cdot J_{n-2}$$
$$= 1 \cdot [1 \cdot J_{n-2} + 2 \cdot J_{n-3}] + 2 \cdot [1 \cdot J_{n-3} + 2 \cdot J_{n-4}]$$
$$= 1 \cdot J_{n-2} + 2 \cdot J_{n-3} + 2 \cdot J_{n-3} + 4 \cdot J_{n-4}$$
$$= 1 \cdot J_{n-2} + 4 \cdot J_{n-3} + 4 \cdot J_{n-4} \qquad \blacksquare$$

## Section 5.2

1. Verify formula (2) for **(a)** n = 5 and **(b)** n = 7.
2. Verify formula (3) for **(a)** n = 7 and **(b)** n = 8.
3. Verify formula (4) for n = 9.
4. Prove formula (3) by using formula (1), the recursive rule for the Jacobsthal sequence.
5. Verify formula (5) for n = 10.
6. Write the string dot product using row 6 of the Duplex Triangle and the the appropriate sequence of consecutive Jacobsthal numbers.

The Jacobsthal recursive formula also fits nicely into another pattern of recursive formulas for $J_n$, see formulas (6) through (10).

$$J_n = 1 \cdot J_{n-1} + 2 \cdot 1 \cdot J_{n-2} = J_2 \cdot J_{n-1} + 2 \cdot J_1 \cdot J_{n-2}, \; n \geq 2 \tag{6}$$

$$= 3 \cdot J_{n-2} + 2 \cdot 1 \cdot J_{n-3} = J_3 \cdot J_{n-2} + 2 \cdot J_2 \cdot J_{n-3}, \; n \geq 3 \tag{7}$$

$$= 5 \cdot J_{n-3} + 2 \cdot 3 \cdot J_{n-4} = J_4 \cdot J_{n-3} + 2 \cdot J_3 \cdot J_{n-4}, \; n \geq 4 \tag{8}$$

$$= 11 \cdot J_{n-4} + 2 \cdot 5 \cdot J_{n-5} = J_5 \cdot J_{n-4} + 2 \cdot J_4 \cdot J_{n-5}, \; n \geq 5 \tag{9}$$

$$= 21 \cdot J_{n-5} + 2 \cdot 11 \cdot J_{n-6} = J_6 \cdot J_{n-5} + 2 \cdot J_5 \cdot J_{n-6}, \; n \geq 6 \tag{10}$$

etc.

Formulas (6) thru (10) are similar to formulas (11) thru (15) for the Pell numbers in Chapter 4, except here the second term in the right member is multiplied by 2.

**Example:** Verify formula (8) for n = 6.
**Solution:** $J_6 = 21$ and

$$
\begin{aligned}
J_4 \cdot J_{n-3} + J_3 \cdot J_{n-4} &= 5 \cdot J_{n-3} + 2 \cdot 3 \cdot J_{n-4} \\
&= 5 \cdot J_3 + 2 \cdot 3 \cdot J_2 \\
&= 5 \cdot 3 + 2 \cdot 3 \cdot 1 \\
&= 15 + 6 \\
&= 21
\end{aligned}
$$

---

7.    Verify formula (8) for n = 5.
8.    Verify formula (9) for n = 7.
9.    Verify formula (10) for n = 6.
10.   Predict the next formula to follow formula (10).

---

We can also prove formula (8) by using formula (6), the recursive rule for the Jacobsthal sequence.

Given Formula (6): $J_n = 1 \cdot J_{n-1} + 2 \cdot 1 \cdot J_{n-2} = J_2 \cdot J_{n-1} + 2 \cdot J_1 \cdot J_{n-2}$
Use the rule and replace $J_{n-1}$ with $1 \cdot J_{n-2} + 2 \cdot 1 \cdot J_{n-3}$.
Therefore,

$$
\begin{aligned}
J_n &= 1 \cdot J_{n-1} + 2 \cdot 1 \cdot J_{n-2} \\
&= 1 \cdot [1 \cdot J_{n-2} + 2 \cdot 1 \cdot J_{n-3}] + 2 \cdot 1 \cdot J_{n-2} \\
&= 1 \cdot J_{n-2} + 2 \cdot J_{n-3} + 2 \cdot 1 \cdot J_{n-2} \\
&= 3 \cdot J_{n-2} + 2 \cdot J_{n-3}
\end{aligned}
$$

Now replace $J_{n-2}$ with $1 \cdot J_{n-3} + 2 \cdot 1 \cdot J_{n-4}$, to get

$$3 \cdot J_{n-2} + 2 \cdot J_{n-3} = 3 \cdot [1 \cdot J_{n-3} + 2 \cdot 1 \cdot J_{n-4}] + 2 \cdot J_{n-3}$$
$$= 3 \cdot J_{n-3} + 6 \cdot J_{n-4} + 2 \cdot J_{n-3}$$
$$= 5 \cdot J_{n-3} + 6 \cdot J_{n-4}$$

Therefore,
$$J_n = 5 \cdot J_{n-3} + 2 \cdot 3 \cdot J_{n-4} = J_4 \cdot J_{n-3} + 2 \cdot J_3 \cdot J_{n-4}, \ n \geq 4 \qquad \blacksquare$$

---

**11.**     Prove formula (7) by using formula (6), the recursive rule for the Jacobsthal sequence.

---

**In general, for given n and j, $j \leq n$,**

$$J_n = J_j \cdot J_{n-(j-1)} + 2 \cdot J_{j-1} \cdot J_{n-j}, \ n \geq j \qquad (11)$$

---

**12.**     Let $j = 7$ and write out the formula (11).
**13.**     Verify formula (11) for $n = 8$ and $j = 7$.
**14.**     Let $n = 2j - 1$ in formula (11) and show that $J_{2j-1} = (J_j)^2 + 2 \cdot (J_{j-1})^2$.

## 5.3 String Dot Products Involving Jacobsthal Numbers

Consider the rth row of the Duplex Triangle (which consists of $r + 1$ elements), then choose $r + 1$ consecutive Jacobsthal numbers from the Jacobsthal sequence. Now form the string dot product of the string corresponding to the rth row of the Duplex Triangle and the string of **r + 1** consecutive Jacobsthal numbers. If the last Jacobsthal number in the string equals $J_k$ (in position k) of the Jacobsthal sequence, then the result of the product equals $J_{r+k}$.

> **Example:** Choose row 3 and $3 + 1 = 4$ consecutive Jacobsthal numbers starting with 3, that is, $(3, 5, 11, 21)$. Then the last number is **21 = $J_6$** and is, therefore, in position 6. Thus, $J_{3+6} = J_9 =$ **171** is the result of the string dot product. The string for the $3^{rd}$ row of the Duplex Triangle is $(8, 12, 6, 1)$. The result of the string dot product is
>
> $$(8, 12, 6, 1) \bullet (3, 5, 11, 21) = 8 \cdot 3 + 12 \cdot 5 + 6 \cdot 11 + 1 \cdot 21$$
> $$= 24 + 60 + 66 + 21$$
> $$= 171 = J_9.$$

     **Section 5.3**

---

**1.**     Form the string of the elements in row 2 of the Duplex Triangle and the string $(5, 11, 21)$ of 3 consecutive elements of the Jacobsthal sequence. Find the string dot product of these two strings and show that it is equal to $J_8 =$ **85**.

2.    Choose row 3 of the Duplex Triangle and $3 + 1 = 4$ consecutive Jacobsthal numbers starting with $J_2 = 1$. Form the string dot product of the string for row 3 of the Duplex Triangle and the string of 4 consecutive Jacobsthal numbers starting with $J_2 = 1$.
Verify that the position of last Jacobsthal number in the string is in position $n = 5$ and that the result of the string dot product is equal to
$$J_{3+5} = J_8 = 85.$$

3.    Form the string consisting of $J_4$ and $J_5$. Use row 1 of the Duplex Triangle to form the second string and show that the string dot product of the two strings is equal to $J_6$.

In Chapter 4, formula (18), we found the string dot product of the string of Fibonacci numbers from the first up to the nth and the string of Pell numbers from the nth down to the first is equal to the difference between the $(n + 1)$th Pell number and the $(n + 1)$th Fibonacci number. That is,

$$(1, 1, 2, 3, 5, \ldots, F_n) \bullet (P_n, P_{n-1}, P_{n-2}, \ldots, 12, 5, 2, 1) = P_{n+1} - F_{n+1}.$$

There is a similar formula involving the Jacobsthal and Fibonacci numbers, but with a small modification in the subscripts of the right-hand member.

$$(1, 1, 2, 3, 5, \ldots, F_n) \bullet (J_n, J_{n-1}, J_{n-2}, \ldots, 5, 3, 1, 1) = J_{n+2} - F_{n+2}. \quad (12)$$

4.    Verify the formula $(1, 1, 2, 3, 5) \bullet (11, 5, 3, 1, 1) = J_7 - F_7$
5.    Verify the formula (12) for $n = 3, 4$ and 6.

The Jacobsthal numbers came from the Jacobsthal diagonals of the Duplex Triangle and we can, therefore, develop an explicit formula for them in terms of the elements in the rows of the Duplex Triangle. Recall, on the $5^{th}$ Jacobsthal diagonal of the Duplex Triangle we have $J_5 = 1 + 6 + 4 = 11$. See the figure on the first page of this chapter. The elements 1, 6, and 4 are equal to $N(4, 0)$, $N(3,1)$ and $N(2, 2)$, respectively. Therefore, $J_5 = N(4, 4) + N(3, 2) + N(2, 0)$. Using the formula $N(d, k) = C(d, k) \cdot 2^{d-k}$, we also have

$$J_5 = N(4, 4) + N(3, 2) + N(2, 0)$$
$$= C(4, 4) \cdot 2^{4-4} + C(3, 2) \cdot 2^{3-2} + C(2, 0) \cdot 2^{2-0}$$
$$= C(4, 4) \cdot 2^0 + C(3, 2) \cdot 2^1 + C(2, 0) \cdot 2^2$$

The last formula expresses $J_5$ in terms of the binomial coefficients of Pascal's Triangle and powers of 2. The specific binomial coefficients are those found on a Fibonacci diagonal. We have the string dot product of the string of binomial coefficients on a Fibonacci diagonal and the string consisting of the powers of 2, in this case the string $(2^0, 2^1, 2^2)$.

$$C(4, 4) \cdot 2^0 + C(3, 2) \cdot 2^1 + C(2, 0) \cdot 2^2 = 1 \cdot 1 + 3 \cdot 2 + 1 \cdot 4 = 1 + 6 + 4 = 11$$

In general, the steps to express $J_n$ in terms of Duplex Triangle elements are as follows:

    1) If n is even divide by 2 to determine the number of terms needed.

       If n is odd, add 1, then divide by 2.

    2) Sum the terms

$$N(n - 1, n - 1) + N(n - 2, n - 3) + N(n - 3, n - 5) + \ldots + N(n - j, n - (2j - 1))$$

until the number of terms equals the value determined in step 1. That happens when $n - (2j - 1) = 0$ or 1.

The general formula is:

$$J_n = N(n - 1, n - 1) + N(n - 2, n - 3) + N(n - 3, n - 5) + \ldots + N(n - j, n - (2j - 1))$$

---

6.      Write the expression for $J_6$ (a) in terms of $N(d, k)$ and (b) in terms of $C(d, k)$ and the powers of 2.

7.      Write the expression for $J_7$ in terms of $N(d, k)$.

8.      (a)  Complete the list below:

$$1 = 1$$
$$2 - 1 = 1$$
$$4 - 2 + 1 = 3$$
$$8 - 4 + 2 - 1 = 5$$
$$16 - 8 + 4 - 2 + 1 = 11$$
$$32 - 16 + 8 - 4 + 2 - 1 = \underline{\quad}$$
$$64 - 32 + 16 - 8 + 4 - 2 + 1 = \underline{\quad}$$

     (b)  Write the next line to fit the pattern above:

9.      Create a string of elements from a Fibonacci diagonal chosen from the figure shown, that matches the length of the given strings below and form the string dot product of the two strings. In each case describe the final result in terms of Jacobsthal numbers.

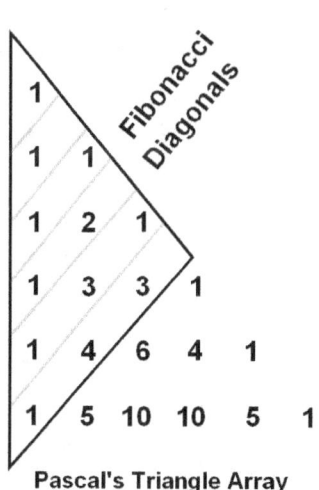

Pascal's Triangle Array

     (a)  $(2^0, 2^1)$

     (b)  $(2^0, 2^1)$ *

     (c)  $(2^0, 2^1, 2^2)$

     (d)  $(2^0, 2^1, 2^2)$ *

     (e)  $(2^0, 2^1, 2^2, 2^3)$

        (Find a new diagonal of length 4.)

     * Use a diagonal different from one previously used.

10. The following string dot products use the rows of the Simplex Triangle for the first string and the powers of –3 forming the second string. Complete the list below, taking the absolute value of the final result:

$$| (3, 3, 1) \bullet (3^0, -3^1, 3^2) | = | 3 - 9 + 9 | = 3 = J_3$$
$$| (4, 6, 4, 1) \bullet (3^0, -3^1, 3^2, -3^3) | = | 4 - 18 + 36 - 27| = -5 = -J_4$$
$$| (5, 10, 10, 5, 1) \bullet (3^0, -3^1, 3^2, -3^3, 3^4) | = | 5 - 30 + 90 - 135 + 81 | = 11 = J_5$$
$$| (6, 15, 20, 15, 6, 1) \bullet (3^0, -3^1, 3^2, -3^3, 3^4, -3^5) | = \underline{\hspace{3cm}}$$

## 5.4 Other Properties of the Jacobsthal Numbers

Like the Pell and Fibonacci numbers, there is a plethora of formulas and relationships for the Jacobsthal numbers. There are other properties of the Jacobsthal numbers that fit into patterns that are only partly similar to patterns we found for the Fibonacci and Pell numbers. For example, recall the formulas for the Fibonacci and Pell numbers:

$$(F_n)^2 - (F_{n-1}) \cdot (F_{n+1}) = \pm 1$$

and

$$(P_n)^2 - (P_{n-1}) \cdot (P_{n+1}) = \pm 1.$$

However, the formula for Jacobsthal numbers is somewhat different

$$(J_n)^2 - (J_{n-1}) \cdot (J_{n+1}) = \pm 1 \cdot 2^{n-1}.$$

Another explicit formula for $J_n$ is

$$J_n = (2^n - (-1)^n) / 3.$$

The next group of exercises illustrate the formulas above and additional formulas involving the Jacobsthal numbers.

### Section 5.4

1. Verify the formula $(J_n)^2 - (J_{n-1}) \cdot (J_{n+1}) = \pm 1 \cdot 2^{n-1}$ for n = 3, 4, and 5.
2. Verify the formula $J_n = (2^n - (-1)^n) / 3$ for n = 3, 4, and 5.
3. Verify the formula $J_{2n+1} + J_{2n} = 12(J_{2n-2}) + 4$ for n = 3, 4, and 5.
4. Verify the formula $(J_{n+3} + J_{n+4})/(J_3 + J_4) = 2^n$ for n = 3, 4, and 5.
5. Verify the formulas (a) $J_{2n} = 2 \cdot J_{2n-1} - 1$ and (b) $J_{2n} = 4 \cdot J_{2n-2} + 1$ and (c) $J_{2n} / J_n = J_{n+1} + 2 \cdot J_{n-1}$ for n = 3, 4, and 5.
6. Verify the formulas (a) $J_{2n+1} = 2 \cdot J_{2n} + 1$ and (b) $J_{2n+1} = 8 \cdot J_{2n-2} + 3$ for n = 3, 4, and 5.
7. Verify the formula $J_n = 2 \cdot J_{n-1} + (-1)^{n-1}$ for n = 3, 4, and 5.
8. Verify the formula $J_n = 4 \cdot J_{n-2} + (-1)^{n-2} \cdot J_2$ for n = 3, 4, and 5.
9. Verify the formula $J_n = 8 \cdot J_{n-3} + (-1)^{n-3} \cdot J_3$ for n = 4, 5, and 6.
10. Verify the formula $J_n = 16 \cdot J_{n-4} + (-1)^{n-4} \cdot J_4$ for n = 5, 6, and 7.

11. Verify the formula $J_n = 2^k \cdot (J_{n-k}) + (-1)^{n-k} \cdot J_k$ for (a) k = 2 and n = 3, 4, and 5 and (b) k = 3 and n = 4, 5, and 6.

12.     Verify the formula $J_{n+1} \cdot J_{n-1} = (J_n)^2 + (-1)^n \cdot 2^{n-1}$ for $n = 3, 4,$ and $5$.

The sums of two Jacobsthal numbers often involve the powers of 2 and other Jacobsthal numbers. Some of these are developed in the next set of exercises.

13.     Verify the formula $J_n + J_{n+1} = 1 \cdot 2^n$ for $n = 3, 4,$ and $5$.
14.     Verify the formula $J_n + J_{n+2} = 1 \cdot 2^n + 2 \cdot J_n$ for $n = 3, 4,$ and $5$.
15.     Verify the formula $J_n + J_{n+3} = 3 \cdot 2^n$ for $n = 3, 4,$ and $5$.
16.     Verify the formula $J_n + J_{n+4} = 5 \cdot 2^n + 2 \cdot J_n$ for $n = 3, 4,$ and $5$.
17.     Verify the formula $J_n + J_{n+5} = 11 \cdot 2^n$ for $n = 3, 4,$ and $5$.
18.     Verify the formula $J_n + J_{n+6} = 21 \cdot 2^n + 2 \cdot J_n$ for $n = 3, 4,$ and $5$.
19.     Verify the formula $J_n + J_{n+7} = 43 \cdot 2^n$ for $n = 3, 4,$ and $5$.
20.     Examine the formulas in Exercises 13 thru 19 and develop the next two formulas in the sequence you observe. Hint: Pay particular attention to the coefficients of the powers of 2.
21.     Propose general formulas for $J_n + J_{n+k}$, one for k equal to an odd number and one for k equal to an even number. Hint: Examine the formulas in Exercises 13 thru 19.
22.     Verify the formula $J_{n+1} + J_{n+1} = (1 + 1) \cdot 2^n - 2 \cdot J_n$ for $n = 3, 4,$ and $5$.
23.     Verify the formula $J_{n+1} + J_{n+2} = (1 + 1) \cdot 2^n$ for $n = 3, 4,$ and $5$.
24.     Verify the formula $J_{n+1} + J_{n+3} = (1 + 3) \cdot 2^n - 2 \cdot J_n$ for $n = 3, 4,$ and $5$.
25.     Verify the formula $J_{n+1} + J_{n+4} = (1 + 5) \cdot 2^n$ for $n = 3, 4,$ and $5$.
26.     Verify the formula $J_{n+1} + J_{n+5} = (1 + 11) \cdot 2^n - 2 \cdot J_n$ for $n = 3, 4,$ and $5$.
27.     Examine the formulas in Exercises 22 thru 26 and develop the next two formulas in the sequence you observe. Hint: Pay particular attention to the coefficients of the powers of 2.
28.     Propose general formulas for $J_{n+1} + J_{n+k}$, $k \geq 2$, one for k equal to an odd number and one for k equal to an even number. Hint: Examine the formulas in Exercises 22 thru 26.
29.     Verify the formula $J_{n+2} + J_{n+2} = (1 + 1) \cdot 2^n + 2 \cdot J_n$ for $n = 3, 4,$ and $5$.
30.     Verify the formula $J_{n+2} + J_{n+3} = (1 + 3) \cdot 2^n$ for $n = 3, 4,$ and $5$.
31.     Verify the formula $J_{n+2} + J_{n+4} = (1 + 5) \cdot 2^n + 2 \cdot J_n$ for $n = 3, 4,$ and $5$.
32.     Verify the formula $J_{n+2} + J_{n+5} = (1 + 11) \cdot 2^n$ for $n = 3, 4,$ and $5$.
33.     Verify the formula $J_{n+2} + J_{n+6} = (1 + 21) \cdot 2^n + 2 \cdot J_n$ for $n = 3, 4,$ and $5$.
34.     Verify the formula $J_{n+2} + J_{n+7} = (1 + 43) \cdot 2^n$ for $n = 3, 4,$ and $5$.
35.     Verify the formula $J_{n+2} + J_{n+8} = (1 + 85) \cdot 2^n + 2 \cdot J_n$ for $n = 3, 4,$ and $5$.
36.     Examine the formulas in Exercises 29 thru 35 and develop the next two formulas in the sequence you observe. Hint: Pay particular attention to the coefficients of the powers of 2.
37.     Propose general formulas for $J_{n+2} + J_{n+k}$, $k \geq 2$, one for k equal to an odd number and one for k equal to an even number. Hint: Examine the formulas in Exercises 29 thru 35.

38.　　Verify the formula $J_{n+3} + J_{n+4} = (3 + 5) \cdot 2^n$ for $n = 3, 4,$ and $5.$

39.　　Verify the formula $J_{n+3} + J_{n+5} = (3 + 11) \cdot 2^n - 2 \cdot J_n$ for $n = 3, 4,$ and $5.$

40.　　Verify the formula $J_{n+3} + J_{n+6} = (3 + 21) \cdot 2^n$ for $n = 3, 4,$ and $5.$

41.　　Verify the formula $J_{n+3} + J_{n+7} = (3 + 43) \cdot 2^n - 2 \cdot J_n$ for $n = 3, 4,$ and $5.$

42.　　Examine the formulas in Exercises 38 thru 41 and develop the next two formulas in the sequence you observe. Hint: Pay particular attention to the coefficients of the powers of 2.

43.　　Propose general formulas for $J_{n+3} + J_{n+k},$ $k \geq 4,$ one for k equal to an odd number and one for k equal to an even number. Hint: Examine the formulas in Exercises 38 thru 41.

Note: We can rewrite the formulas for Exercises 18 and 19 <u>and</u> 34 and 35 as follows:

$$J_n + J_{n+6} = 21 \cdot 2^n + 2 \cdot J_n = J_6 \cdot 2^n + 2 \cdot J_n$$

$$J_n + J_{n+7} = 43 \cdot 2^n = J_7 \cdot 2^n$$

and

$$J_{n+2} + J_{n+7} = (1 + 43) \cdot 2^n = (J_2 + J_7) \cdot 2^n$$

$$J_{n+2} + J_{n+8} = (1 + 85) \cdot 2^n + 2 \cdot J_n = (J_2 + J_8) \cdot 2^n + 2 \cdot J_n$$

44.　　Rewrite the formulas for Exercises 16 and 17 <u>and</u> 32 and 33, making the substitutions for the coefficients of $2^n$ in terms of $J_{n+k},$ choosing the appropriate values for k.

45.　　Rewrite the formulas for Exercises 40 and 41, making the substitutions for the coefficients of $2^n$ in terms of $J_{n+k},$ choosing the appropriate values for k.

46.　　Verify the formula $J_{n+k} = J_k \cdot 2^n + J_n$ for $k = 4$ and $n = 2, 4,$ and $6.$

47.　　Verify the formula $J_{n+k} = J_k \cdot 2^n - J_n$ for $k = 5$ and $n = 1, 3,$ and $5.$

When $J_{2n}$ is expressed in binary form, an interesting pattern develops. For example, expressing $J_4 = 5$ and $J_6 = 21$ in binary, we get **101** and **10101**, respectively.

48.　　Express $J_8 = 85$ and $J_{10} = 341$ in binary form. Describe the pattern of binary digits. Hint: $85 = 64 + 16 + 4 + 1$ and $341 = 256 + 64 + 16 + 4 + 1.$

## 5.5 The Companion Jacobsthal Number Sequence

The **Companion Jacobsthal number sequence**, sometimes called the **Jacobsthal-Lucas** sequence, is an infinite sequence of integers that conforms to the same recursive formula as the Jacobsthal sequence, but uses different starting values for the first two elements in the sequence. The companion sequence is

$$2, 1, 5, 7, 17, 31, 65, 127, 257, 511, 1025, \ldots, K_n, \ldots$$

where $n = 0, 1, 2, 3, \ldots$, and $K_n$ represents the nth number in the sequence and follows the recursive rule

$$K_n = K_{n-1} + 2 \cdot K_{n-2}$$

for $n = 2, 3, \ldots$ and $K_0 = 2$ and $K_1 = 1$.

For example, $K_5 = K_4 + 2 \cdot K_3$, or equivalently, $31 = 17 + 2 \cdot 7$.

Many of the formulas valid for the Jacobsthal numbers are also valid for the Companion Jacobsthal numbers. For instance, the recursive rule for the Jacobsthal-Lucas numbers is formed from the <u>string dot product</u> of the <u>reverse</u> of row 1 of the Duplex Triangle and two previous consecutive elements of the Jacobsthal-Lucas sequence, giving us

$$K_n = (1, 2) \bullet (K_{n-1}, K_{n-2}) = 1 \cdot K_{n-1} + 2 \cdot K_{n-2}$$

In general, we have the following formulas, which are the Duplex extensions of the recursive Jacobsthal-Lucas Rule.

$$
\begin{aligned}
K_n = \ &1 \cdot K_n \\
= \ &1 \cdot K_{n-1} + 2 \cdot K_{n-2} &(13)\\
= \ &1 \cdot K_{n-2} + 4 \cdot K_{n-3} + 4 \cdot K_{n-4} &(14)\\
= \ &1 \cdot K_{n-3} + 6 \cdot K_{n-4} + 12 \cdot K_{n-5} + 8 \cdot K_{n-6} &(15)\\
= \ &1 \cdot K_{n-4} + 8 \cdot K_{n-5} + 24 \cdot K_{n-6} + 32 \cdot K_{n-7} + 16 \cdot K_{n-8} &(16)\\
= \ &1 \cdot K_{n-5} + 10 \cdot K_{n-6} + 40 \cdot K_{n-7} + 80 \cdot K_{n-8} + 80 \cdot K_{n-9} + 32 \cdot K_{n-10} &(17)
\end{aligned}
$$

etc.

The formulas above are analogous to formulas (1) through (5) for the Jacobsthal sequence.

There are formulas that relate the Jacobsthal numbers to the Companion Jacobsthal numbers. For instance, the product of the nth Jacobsthal number and the nth Jacobsthal-Lucas number is equal to the (2n)th Jacobsthal number. The formula is

$$J_n \cdot K_n = J_{2n}. \qquad (18)$$

Formula (18) is similar to the formulas we found previously for the Pell and Fibonacci numbers:

$$P_n \cdot Q_n = P_{2n} \quad \text{and} \quad F_n \cdot L_n = F_{2n}.$$

An explicit formula for $K_n$ is

$$K_n = 2^n + (-1)^n.$$

Other formulas are

$$K_n = J_{n+1} + 2 \cdot J_{n-1},$$

$$J_n + K_n = 2 \cdot J_{n+1},$$

$$3 \cdot J_n + K_n = 2^{n+1},$$

$$(K_n)^2 - (K_{n-1}) \cdot (K_{n+1}) = (-1)^n \cdot 9 \cdot 2^{n-1},$$

$$(K_{n+1})^2 + 2 \cdot (K_n)^2 = 3 \cdot (K_{2n+1} + 2),$$

and
$$K_{n+1} = 2 \cdot K_n - (-1)^n \cdot 3.$$

### Section 5.5

1.  Verify the formula $K_n = 4 \cdot K_{n-2} + 4 \cdot K_{n-3} + 1 \cdot K_{n-4}$ for $n = 5$.
2.  Verify the formula $J_n \cdot K_n = J_{2n}$ for $n = 3, 4$, and 5.
3.  Verify the formula $K_n = J_{n+1} + 2 \cdot J_{n-1}$ for $n = 3, 4$, and 5.
4.  Verify the formula $J_n + K_n = 2 \cdot J_{n+1}$ for $n = 3, 4$, and 5.
5.  Verify the formula $3 \cdot J_n + K_n = 2^{n+1}$ for $n = 3, 4$, and 5.
6.  Verify the formula $K_n = 2^n + (-1)^n$ for $n = 3, 4$, and 5.
7.  Verify the formula $K_{n+1} = 2 \cdot K_n - (-1)^n \cdot 3$ for $n = 3, 4$, and 5.
8.  Verify the formulas $(K_n)^2 - (K_{n-1}) \cdot (K_{n+1}) = (-1)^n \cdot 9 \cdot 2^{n-1}$
    and $(K_{n+1})^2 + 2 \cdot (K_n)^2 = 3 \cdot (K_{2n+1} + 2)$ for $n = 3, 4,$ and 5.

There is an interesting connection between the Duplex Triangle and the Jacobsthal-Lucas numbers.   Form the string consisting of the elements of the 3rd row of the Duplex Triangle and the string of the Jacobsthal-Lucas numbers from the first to the fourth (there are 4 elements in the 3rd row of the Duplex Triangle).  Then find the <u>string dot product</u> to get the 7th Jacobsthal-Lucas number.

$$K_7 = (8, 12, 6, 1) \bullet (1, 5, 7, 17)$$

$$= 8 + 60 + 42 + 17 = 127$$

We find this to be the Jacobsthal-Lucas number in the 7th position of the Jacobsthal-Lucas sequence by adding the row number of the Duplex Triangle used in the first string to the position number in the Jacobsthal-Lucas sequence of the last Jacobsthal-Lucas number in the second string.  Since we used row 3 and the position number of 17 is 4 in the sequence, we have, $3 + 4 = 7$, thus the result of the string dot product is $K_7$.

*In general, choose row r of the Duplex Triangle for the first string.*
*Then choose a string of (r + 1) consecutive Jacobsthal-Lucas numbers,*
*noting the index of the largest or last Lucas number in the second string,*
*say k.  Then the string dot product of these two strings equals the (r + k)th*
*Jacobsthal-Lucas number.*

**Example:** Choose row 2 of the Duplex Triangle. Then choose $2 + 1 = 3$ consecutive Jacobsthal-Lucas numbers starting with 5, that is, (5, 7, 17). The string dot product is

$$(4, 4, 1) \bullet (5, 7, 17) = 20 + 28 + 17 = 65 = K_6$$

The subscript 6 is found by adding 2 (the row number) to the index position of the last Jacobsthal-Lucas number in the second string which is 4, since $17 = K_4$. Thus, $2 + 4 = 6$.

9.    Find the string dot product of the string consisting of the $3^{rd}$ row of the Duplex Triangle and 4 consecutive Jacobsthal-Lucas numbers starting with $5 = K_2$. Verify that the result is $K_8$.

10.   Choose your own row of the Duplex Triangle and your own string of consecutive Jacobsthal-Lucas numbers consistent with the length of the first string corresponding to the row you chose. Find the Jacobsthal-Lucas number and its index that equals the string dot product of your two strings.

## 5.6   The Sum of the Jacobsthal Numbers

The sum of Jacobsthal numbers from the first to the nth, $1 + 1 + 3 + 5 + 11 + 21 + \ldots + J_n$, is equal to one-half the quantity of $J_{n+2}$ minus 1. The formula is

$$1 + 1 + 3 + 5 + 11 + 21 + \ldots + J_n = (J_{n+2} - 1)/2 \qquad (19)$$

Another formula is as follows: if n is even, then the sum is equal to $J_{n+1} - 1$ and if n is odd, then the sum is equal to $J_{n+1}$. We can write it this way:

$$1 + 1 + 3 + 5 + 11 + 21 + \ldots + J_{2n} = J_{2n+1} - 1, \text{ and} \qquad (20)$$

$$1 + 1 + 3 + 5 + 11 + 21 + \ldots + J_{2n+1} = J_{2n+2}. \qquad (21)$$

Chapter 4, Formula (22), gives the sum of the Pell numbers as

$$1 + 2 + 5 + 12 + 29 + \ldots + P_n = (P_{n+1} + P_n - 1)/2.$$

By way of comparison, the sum of the Jacobsthal numbers is given above as $(J_{n+2} - 1)/2$, but by the recursive rule, $J_{n+2}$ is equal to the sum of the previous Jacobsthal number plus 2 times the Jacobsthal number prior to that. Therefore, we can write

$$1 + 1 + 3 + 5 + 11 + 21 + \ldots + J_n = (J_{n+2} - 1)/2$$
$$= (J_{n+1} + 2 \bullet J_n - 1)/2. \qquad (22)$$

### Section 5.6

1.  Use the two formulas (19) and (22) to find the sum  $1 + 1 + 3 + 5 + 11 + 21$. Verify that each formula yields the same result.
2.  Find the sum  $1 + 1 + 3 + 5 + 11 + \ldots + J_n$  for $n = 3, 5, 6$ and $7$ using one of the formulas given on the previous page.
3.  Find the sum  $1 + 1 + 3 + 5 + 11 + \ldots + J_n$  for $n = 7$ and $8$ using formulas  (20) and (21).

The sequence of partial sums of the Jacobsthal numbers is also a recursive sequence:

$$1, 2, 5, 10, 21, 42, 85, \ldots, H_n, \ldots$$

where
$$H_n = H_{n-1} + 2 \cdot H_{n-2} + 1$$

and  $H_1 = 1$ and $H_2 = 2$  and $n = 3, 4, 5, \ldots$.

4.  Verify the formula $H_n = H_{n-1} + 2 \cdot H_{n-2} + 1$  for $n = 3, 4,$ and $5$.
5.  Verify the formula $H_n = (J_{n+1} + 2 \cdot J_n - 1)/2$  for $n = 3, 4,$ and $5$.
6.  Verify the formula $2^n = H_{n-1} + H_n + 1$  for $n = 3, 4,$ and $5$.

The sum of the Jacobsthal-Lucas numbers is given by the formula

$$1 + 5 + 7 + 17 + 31 + 65 + 127 + 257 + \ldots + K_n = (K_{n+2} - 5)/2 . \qquad (23)$$

Another formula is as follows:  if n is even, then the sum is equal to  $K_{n+1} - 1$ and if n is odd, then the sum is equal to  $K_{n+1} - 4$. We can write it this way:

$$1 + 5 + 7 + 17 + 31 + 65 + \ldots + K_{2n} = K_{2n+1} - 1, \text{ and} \qquad (24)$$

$$1 + 5 + 7 + 17 + 31 + 65 + \ldots + K_{2n+1} = K_{2n+2} - 4 . \qquad (25)$$

7.  Use the formulas, (23) and (24) or (25), to find the sum $1 + 5 + 7 + 17 + 31$ .  Verify that each formula yields the same result.
8.  Use the formulas, (23) and (24) or (25), to find the sum $1 + 5 + 7 + 17 + 31 + 65$ .  Verify that each formula yields the same result.

## 5.7  Related Sequences of the Jacobsthal Numbers

We have a relationship with the  Duplex Triangle  and the Jacobsthal numbers  similar to the relationship we explored with the Duplex Triangle and the Pell diagonals in Chapter 4. We first look at the sum of Jacobsthal numbers from $J_1$ to $J_n$ as given by formula (19), and that sum is equal to

$$(J_{n+2} - 1)/2.$$

Also, the sum of all the elements in the Duplex Triangle from row 0 to row $n - 1$ is equal to

$$(1/2)(3^n - 1)$$

**Duplex Triangle Array**

(refer to Chapter 3).

Since the sum of the Jacobsthal numbers, as found on the Jacobsthal diagonals in the Duplex Triangle, correspond to elements enclosed by the triangle shown, the formula for the sum of the Duplex Triangle elements **not** represented by numbers on the Jacobsthal diagonals for a given value of n, is equal to the sum of all the elements of the Duplex Triangle up to row ($n -$ 1) minus the sum of the Jacobsthal numbers from $J_1$ to $J_n$. That is,

$$(1/2)(3^n - 1) - (J_{n+2} - 1)/2 = (3^n - J_{n+2})/2 .$$

Since $J_n = (2^n - (-1)^n)/3,$ we can substitute $(2^{n+2} - (-1)^{n+2})/3 = (2^{n+2} - (-1)^n)/3$ for $J_{n+2}$ in the formula above. That gives us the sum of the Duplex Triangle elements through row n - 1 **not** represented by numbers on the Jacobsthal diagonals for a given value of n, is equal to

$$(3^n - J_{n+2})/2 = [3^n - (2^{n+2} - (-1)^n)/3 ]/2$$
$$= (3^{n+1} - 2^{n+2} + (-1)^n)/6 .$$

**Section 5.7**

1.    Evaluate the two formulas $(3^n - J_{n+2})/2$ and $(3^{n+1} - 2^{n+2} + (-1)^n)/6$ for n = 3, 4 and 5. Verify that each formula yields the same result as the other.

2.    For each of the figures on the next page, find
   **(a)** the sum of all of the elements shown in the Duplex Triangle,
   **(b)** the sum of the elements enclosed in the drawn triangle, and
   **(c)** the sum of the elements not enclosed in the triangle.
   In each case, compare your sums to the values of the expressions

   (1)   $(1/2)(3^n - 1)$,                       (2)   $(J_{n+2} - 1)/2,$

   (3)   $(1/2)(3^n - 1) - (J_{n+2} - 1)/2,$   and   (4)   $(3^{n+1} - 2^{n+2} + (-1)^n)/6 .$

   where n equals one more than the row number of the last row shown in each figure.

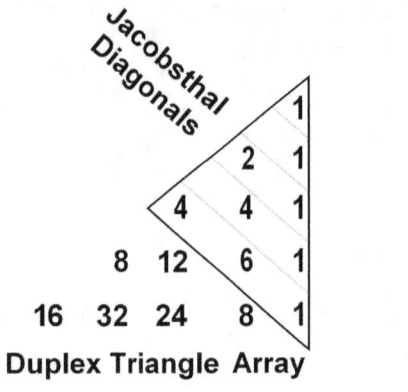

Duplex Triangle Array

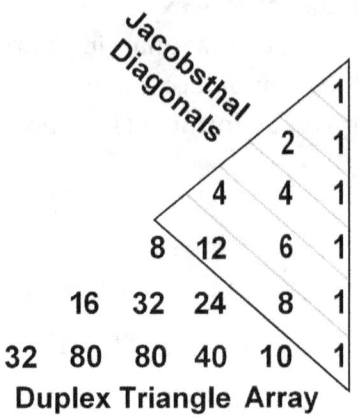

Duplex Triangle Array

3. Refer to the figure at the beginning of this section and find the number of the last row shown. (a) Then choose a formula from the previous page to find the sum of the elements not enclosed in the triangle. (b) Sum the elements directly that are shown in the figure that are not enclosed in the triangle and compare the total to your answer in part (a).

The expression

$$(1/2)(3^n - 1) - (J_{n+2} - 1)/2 = (3^n - J_{n+2})/2,$$

given for the sum of those elements of the Duplex Triangle that are **not enclosed** with the Jacobsthal diagonals, provides a formula for a sequence of numbers which we will denote by

$$c(n) = (3^n - J_{n+2})/2, \quad \text{for } n = 1, 2, 3, \ldots . \tag{26}$$

The formula for **c(n)** produces the sequence

**0, 2, 8, 30, 100, 322, 1008, 3110, . . ., c(n), . . . .**

Also, notice that $c(4) = 30 = 3 \cdot c(3) + 2 \cdot 3$ and $c(5) = 100 = 3 \cdot c(4) + 2 \cdot 5$. The 3 and 5 that were multiplied by 2 and added in those two formulas are the Jacobsthal numbers $J_3$ and $J_4$. Noting this we can write $c(4) = 3 \cdot c(3) + 2 \cdot J_3$ and $c(5) = 3 \cdot c(4) + 2 \cdot J_4$.

4. Complete the table on the next page. Compare the final values of columns 3 and 4 with the values in column 2. Make a conjecture.

In general,

$$c(n) = (3^n - J_{n+2})/2 = 3 \cdot c(n-1) + 2 \cdot J_{n-1}.$$

| n | c(n) | $(3^n - J_{n+2})/2$ | $3 \cdot c(n-1) + 2 \cdot J_{n-1}$ |
|---|------|---------------------|-----------------------------------|
| 1 | 0 | $(3^1 - J_{1+2})/2 =$   0 | $3 \cdot c(0) + 2 \cdot J_0 =$   0 |
| 2 | 2 | $(3^2 - J_{2+2})/2 =$   2 | $3 \cdot c(1) + 2 \cdot J_1 =$   2 |
| 3 | 8 | $(3^3 - J_{3+2})/2 =$   8 | $3 \cdot c(2) + 2 \cdot J_2 =$   8 |
| 4 | 30 | $(3^4 - J_{4+2})/2 =$   30 | $3 \cdot c(3) + 2 \cdot J_3 =$   30 |
| 5 | 100 | $(3^5 - J_{5+2})/2 =$ \_\_\_\_ | $3 \cdot c(4) + 2 \cdot J_4 =$ \_\_\_\_ |
| 6 | 322 | $(3^6 - J_{6+2})/2 =$ \_\_\_\_ | $3 \cdot c(5) + 2 \cdot J_5 =$ \_\_\_\_ |
| 7 | 1008 | _____ $=$ \_\_\_\_ | _____ $=$ \_\_\_\_ |
| 8 | \_\_\_\_ | _____ $=$ \_\_\_\_ | _____ $=$ \_\_\_\_ |

Another formula for the general term of the sequence **c(n)** is

$$c(n) = (3^{n+1} - 2^{n+2} + (-1)^n)/6 .$$

Recall formulas (23) from Chapter 4 for **a(n)**, the sequence associated with Pascal's Triangle and the Fibonacci diagonals, and formula (24) for **b(n)**, the sequence associated with the Duplex Triangle and the Pell diagonals. Compare to the formula for **c(n)**.

$$a(n) = 2 \cdot a(n-1) + F_{n-1}$$
$$b(n) = 3 \cdot b(n-1) + P_{n-1}$$
$$c(n) = 3 \cdot c(n-1) + 2 \cdot J_{n-1} .$$

The formula (26), **c(n)** $= (3^n - J_{n+2})/2$, given for the sum of those elements of the Duplex Triangle that are **not enclosed** with the Jacobsthal diagonals as shown in the figures, provides a formula for a sequence of numbers which we will denote by

$$d(n) = 2 \cdot c(n) = 3^n - J_{n+2}, \quad \text{for } n = 1, 2, 3, \ldots . \tag{27}$$

This produces the sequence

$$\textbf{0, 4, 16, 60, 200, 644, 2016, 6220, } \ldots \textbf{, } 2 \cdot c(n), \ldots,$$

where each element, for a given n, is two times the element in the sequence for **c(n)**.

The sequence **d(n)** turns out to have a interesting connection with a special "hypothetical coin" tossing problem. This hypothetical coin can come up one of three equally likely values when tossed. We will represent these outcomes as {0, 1, 2.}. Suppose you toss this "coin" n times, creating a string of length n consisting of 0's, 1's and 2's. There are $3^n$ possible patterns. How many times will the patterns of 0's, 1's and 2's being generated, show a pattern where *at least two consecutive digits are nonzero*?

For example, suppose you toss this "coin" 3 times. Four of these patterns are as follows: **011, 012, 021, 022.** How many patterns are there? Formula (27), yields the answer, that is, the value of **d(n)** $= 3^n - J_{n+2}$, for n = 3,

$$d(3) = 3^3 - J_{3+2} = 27 - J_5 = 27 - 11 = 16.$$

We will list these 16 patterns below.  Note:  There are only three ways the coin can come up on a single toss, either 0, 1, or 2, therefore, there are  **3 x 3 x 3  = 3³ = 27**  possible patterns. It is useful to represent the sequence of tosses with ternary number patterns.   The 3 outcomes on each trial range from 000 to 222 as ternary numbers.  The list below shows the 16 patterns where at least two consecutive digits are nonzero in the pattern:

> **011,  012,  021,  022,**
> **110,  111,  112,  120, 121, 122,**
> **210,  211,  212,  220, 221, 222.**

Furthermore, since there are only $3^n$ total patterns for tossing this "coin" n times, the expression $J_{n+2}$ in the formula, represents the negation of the requirement of "showing a pattern where at least two consecutive digits are nonzero".   Hence, in our example,  $J_5 = 11$ is the number of patterns with 3 tosses that **do not** show a pattern where two consecutive digits are nonzero.  In our example those patterns are as follows:

> **000, 001, 002, 010, 020, 100, 101, 102, 200, 201, 202** .

*In general, $J_{n+2}$ is equal to the number of patterns with n tosses of the special "coin" that <u>do not</u> show a pattern where two consecutive digits are nonzero.*

---

5.    Suppose you toss the special "coin" 4 times.  How many of the $3^4$ possible outcomes will show a pattern where at least two consecutive digits are nonzero?
How many patterns will <u>not</u> show a pattern where two consecutive digits are nonzero?

6.    Suppose you toss the special "coin" 5 times.  How many of the $3^5$ possible outcomes will show a pattern where at least two consecutive digits are nonzero?
How many patterns will <u>not</u> show a pattern where two consecutive digits are nonzero?

---

The following demonstrates a method to derive the value of **d(n)** from **d(n – 1)** and **d(n – 2)** in a recursive manner .   We will use n = 4 in the demonstration.  First create two lists, one for the outcomes of tossing the "coin" 2 times (that is n – 2) and one for tossing it 3 times (that is n – 1).  (When you learn the method used, it will not be necessary to make full lists.)  The patterns listed below and on the next page assume the numerical order of the base three representations of the outcomes.

**2 tosses** ($3^2 = 9$ patterns):    00, 01, 02,        10, <u>11</u>, <u>12</u>,        20, <u>21</u>, <u>22</u>        ⟶

**3 tosses** ($3^3 = 27$ patterns from 000 to 222 in ternary numerical order):

<u>first part:</u>     000, 001, 002,    010, 0<u>11</u>, 0<u>12</u>,    020, 0<u>21</u>, 0<u>22</u>

<u>second part:</u>    100, 101, 102,   [ 110, 111, 112,    120, 121, 122]

<u>third part:</u>     200, 201, 202,   [ 210, 211, 212,    220, 221, 222]

We can see that there are 4 underlined entries in the set of 9 patterns for 2 tosses that satisfy the the condition *"at least two consecutive digits are nonzero"*, and so **d(2) = 4** . Also, take note that of the 9 patterns for 2 tosses, two-thirds of those patterns start with either a 1 or a 2. This will be important when analyzing the 27 patterns for 3 tosses.

The 27 patterns are divided into 3 parts where each part consists of the patterns for 2 tosses (shown in bold face), and each of those outcomes is preceded by 0 in the first part, 1 in the second part, and 2 in the third part.

In the first part of the 27 patterns for 3 tosses, there are also 4 patterns that satisfy the condition. These are the same 4 patterns found in the scenario of 2 tosses, but now each pattern is preceded by a 0. In the second part, two-thirds, or 6, of the patterns are starting with a 11 or 12 and for those last 6 patterns out of the 9, each pattern will satisfy the condition when prefixed with a 1 (listed in brackets). The same is true for the third part, 6 patterns are prefixed with a 2 and they also satisfy the condition. Therefore, for 3 tosses we have $4 + 6 + 6 = 16$ patterns that satisfy the condition, and **d(3) = 16** .

Now consider 4 tosses ($3^4 = 81$ patterns). Divide the 81 patterns into 3 parts of 27 patterns each. In the first part we see the 27 patterns found for 3 tosses, with each of those outcomes preceded by 0.

**4 tosses** ($3^4 = 81$ patterns from 0000 to 2222 in ternary numerical order):

<u>first part:</u>    0000, 0001, 0002,   0010, 00<u>11</u>, 00<u>12</u>,    0020, 00<u>21</u>, 00<u>22</u>

               0100, 1001, 0102,   0<u>110</u>, 0<u>111</u>, 0<u>112</u>,    0<u>120</u>, 0<u>121</u>, 0<u>122</u>

               0200, 0201, 0202,   0<u>210</u>, 0<u>211</u>, 0<u>212</u>,    0<u>220</u>, 0<u>221</u>, 0<u>222</u>

We can see that there are 16 underlined entries in the set of 27 patterns for 3 tosses that satisfy the the condition, since **d(3) = 16**.

In the second part we see 27 more patterns matching the patterns for 3 tosses, but now with each of those outcomes preceded by a 1.

<u>second part:</u>    1000, 1001, 1002,   1010, 10<u>11</u>, 10<u>12</u>,    1020, 10<u>21</u>, 10<u>22</u>

               [ 1100, 1101, 1102,   1110, 1111, 1112,    1120, 1121, 1122

               1200, 1201, 1202,   1210, 1211, 1212,    1220, 12<u>21</u>, 1222 ]

In the first group of 9 patterns of the second part, only 4 patterns, **d(2) = 4**, satisfy the condition. And for the second and third groups of 9 patterns each (listed in brackets), all start with a 1, and all 18 patterns satisfy the condition. Thus, for the second part $18 + 4 = 22$ patterns out of the 27 satisfy the condition.

The same is true for the <u>third part</u>, 22 patterns, all starting with 2,  also satisfy the condition.  Altogether, for the three parts, there are $16 + 22 + 22 = 60$ patterns for 4 tosses that satisfy the condition, and so **d(4) = 60** .

To summarize,  consider 4 tosses of the "coin", generating $3^4 = 81$  patterns.  To find **d(4)**, we add **d(3) = 16,** found for the first part, plus $(2/3) \cdot 27 = 18$ plus **d(2) = 4** for each of the second and third parts.  That is, $2 \cdot (18 + 4) = 44$  for the second and third parts.  The total is $16 + 44 = 60$, or in terms of the sequence **d(n)**,

$$d(4) = d(3) + 2 \cdot [(2/3) \cdot 3^3 + d(2)]$$

$$= 16 + 2 \cdot [18 + 4] = 60.$$

In general, the constructive recursive formula is

$$d(n) = d(n-1) + 2 \cdot [(2/3) \cdot 3^{n-1} + d(n-2)] \tag{28}$$

---

7.      Consider 5 tosses of the "hypothetical coin"that can come up one of three equally likely values when tossed, {0, 1, 2}.  How many different patterns could be generated?   Given that **d(4) = 60**  and **d(3) = 16,**  use formula (28) to find the number of those patterns that satisfy the condition *"at least two consecutive digits are nonzero"* ?  How many of the patterns **do not** show a pattern where two consecutive digits are nonzero?

8.      Evaluate  formula (28) for $n = 6$.  Verify that the formula yields the sixth element of the sequence
        **0, 4, 16, 60, 200, 644, 2016, 6220, . . . , d(n), . . . .**

9.      Use formula (27), **d(n)** $= 3^n - J_{n+2}$, and substitute $3^n - J_{n+2}$ for **d(n)**, and $3^{n-1} - J_{n+1}$ for **d(n − 1)**, and $3^{n-2} - J_n$ for **d(n − 2)**,  in formula (28) to establish that both members of formula (28) are identical.

---

Recall the recursive formula for **c(n)** given earlier **c(n)** $= 3 \cdot c(n-1) + 2 \cdot J_{n-1}$. There is a similar recursive formula for **d(n)**,

$$d(n) = 3 \cdot d(n-1) + 4 \cdot J_{n-1}.$$

Another formula,  explicit formula, for the general term of the sequence **d(n)**  is

$$d(n) = (3^{n+1} - 2^{n+2} + (-1)^n)/3 .$$

---

10.      Evaluate the formula **d(n)** $= 3 \cdot d(n-1) + 4 \cdot J_{n-1}$ for $n = 3, 4$, and 5. Verify that the formula yields the third, fourth, and fifth element of the sequence  **0, 4, 16, 60, 200, 644, 2016, 6220, . . . , d(n), . . . .**

11.      Evaluate the formula $d(n) = (3^{n+1} - 2^{n+2} + (-1)^n)/3$ for $n = 3$, 4, and 5. Verify that the formula yields the third, fourth, and fifth element of the sequence    **0, 4, 16, 60, 200, 644, 2016, 6220, . . . , d(n), . . . .**

## 5.8 The Jacobsthal and the Jacobsthal-Lucas Pythagorean Connection

In Exercise 16, Section 5.2, we showed that $J_{2j-1} = (J_j)^2 + 2 \cdot (J_{j-1})^2$, or equivalently,

$$J_{2j-1} - (J_{j-1})^2 = (J_j)^2 + (J_{j-1})^2.$$

That is, the sum of the squares of two consecutive Jacobsthal numbers is equal to

$$J_{2j-1} - (J_{j-1})^2.$$

In Section 4.2 we discussed Pythagorean triples in connection with the Fibonacci and Pell sequences of numbers. The sides of the right triangle, **(a, b, c)**, are determined by **Euclid's formulas:**

$$a = x^2 - y^2, \quad b = 2xy, \quad \text{and} \quad c = x^2 + y^2$$

where $x > y$ and **a** and **b** are the legs of the right triangle and **c** is the hypotenuse, so that

$$c^2 = a^2 + b^2.$$

Now if we choose consecutive Jacobsthal numbers for x and y, then c, the hypotenuse, will equal $J_{2j-1} - (J_{j-1})^2$. Examine the table below:

| x | y | $a = x^2 - y^2$ | $b = 2xy$ | $c = x^2 + y^2$ |
|---|---|---|---|---|
| $J_2 = 1$ | 1 | 0 | 2 | $2 = J_3 - 1^2$ — **not a triangle** |
| $J_3 = 3$ | 1 | 8 | 6 | $10 = J_5 - 1^2$ |
| $J_4 = 5$ | 3 | 16 | 30 | $34 = J_7 - 3^2$ |
| $J_5 = 11$ | 5 | 96 | 110 | $146 = J_9 - 5^2$ |
| $J_6 = 21$ | 11 | 320 | 462 | $562 = J_{11} - 11^2$ |
| . . . | . . . | | | |

Notice that the subscripts of the Jacobsthal numbers in the last column are odd numbers and are equal to the sum of the subscripts of the Jacobsthal numbers represented by x and y. Furthermore, the numbers being squared in the last column are Jacobsthal numbers represented by **y** in the second column. That is because

$$c = x^2 + y^2 = (J_{n+1})^2 + (J_n)^2 = J_{2n+1} - (J_n)^2,$$

and this implies

$$J_{2n+1} = (J_{n+1})^2 + 2 \cdot (J_n)^2.$$

Note that the difference, $c - b$, is the square of $2 \cdot (J_{n-1})$, or, $4 \cdot (J_{n-1})^2$. Also note that the sum, $c + b$, is equal to an even power of 2.   That is because

$$c + b = x^2 + 2xy + y^2 = (x + y)^2 = (J_{n+1} + J_n)^2 = (2^n)^2 = 2^{2n}.$$

(Note:  $J_{n+1} + J_n = 2^n$ because of the property presented in Exercise 13 of Section 5.4.)

### Section 5.8

1.  Extend the table on the previous page by forming the next line in the sequence.
2.  Verify that $J_{2n+1} - (J_n)^2$ is equal to the value of $c$ in your extension of the table.
3.  Verify that the difference, $c - b$, in each line of the table is the square of two times a Jacobsthal number.
4.  Verify that the sum,  $c + b$, in each line of the table  is an even power of 2.
5.  For each row of the table of Pythagorean triples on the previous page, show that the triangle perimeter, $a + b + c$,  for that row is equal to the value of $b$ in the next row minus the value of $b$ in the given row.
6.  For each row of the table show that the triangle perimeter, $a + b + c$,  for that row is equal to $(2^{n+1}) \cdot (J_{n+1})$.
7.  For each row of the table of Pythagorean triples on the previous page, show that the area of the triangle, $(a \cdot b)/2$,  for that row is equal to
    (a)  the product $(2^{n+1}) \cdot (J_{n+1}) \cdot (J_n) \cdot (J_{n-1})$, and
    (b)  $(J_n) \cdot (J_{n-1}) \cdot$ Perimeter.

In a table of Pythagorean Triples using consecutive Companion Jacobsthal, or Jacobsthal-Lucas, numbers for the values of $x$ and $y$, we find many relationships concerning the sides, hypotenuse, area and perimeter, similar to the relationships explored in the exercises above for the Jacobsthal numbers.  For example, the perimeter of the triangle for a given row in the table is equal to  the value of $b$ in the next row minus the value of $b$ in the given row.   Furthermore, the perimeter is equal to $3 \cdot (2^{n+1}) \cdot (K_{n+1})$, three times a similar formula for the perimeter given above for the Jacobsthal numbers.   The ratio of the area to the perimeter of the Jacobsthal numbers is $(J_n) \cdot (J_{n-1})$, and is similar to the corresponding ratio for the Jacobsthal-Lucas numbers, which is  $(K_n) \cdot (K_{n-1})$.

8.  Complete the table on the next page for Pythagorean Triples using consecutive Companion Jacobsthal, or Jacobsthal-Lucas,  numbers as inputs for $x$ and $y$.
9.  Verify that the sum,  $c + b$, in each line of the table,  is equal to $3^2 \cdot (2^{2n})$.

| | x | y | $a = x^2 - y^2$ | $b = 2xy$ | $c = x^2 + y^2$ |
|---|---|---|---|---|---|
| $K_2 =$ | 5 | 1 | _____ | _____ | _____ |
| $K_3 =$ | 7 | 5 | _____ | _____ | _____ |
| $K_4 =$ | 17 | 7 | _____ | _____ | _____ |
| $K_5 =$ | 31 | 17 | _____ | _____ | _____ |
| $K_6 =$ | 65 | 31 | _____ | _____ | _____ |
| | . . . | . . . | | | |

10.    For each row of the table of Pythagorean triples above, show that the triangle perimeter, **a + b + c,** for that row is equal to the value of **b** in the next row minus the value of **b** in the given row.

11.    For each row of the table above show that the triangle perimeter, **a + b + c,** for that row is equal to $3 \cdot (2^{n+1}) \cdot (K_{n+1})$.

12.    For each row of the table of Pythagorean triples above, show that the area of the triangle, **(a· b)/2,** for that row is equal to
    (a)   the product $3 \cdot (2^{n+1}) \cdot (K_{n+1}) \cdot (K_n) \cdot (K_{n-1})$, and
    (b)   $(K_n) \cdot (K_{n-1}) \cdot$ Perimeter, and
    (c)   $[(K_{n+3}) - (K_{n+1})] \cdot (K_{n+1}) \cdot (K_n) \cdot (K_{n-1})$

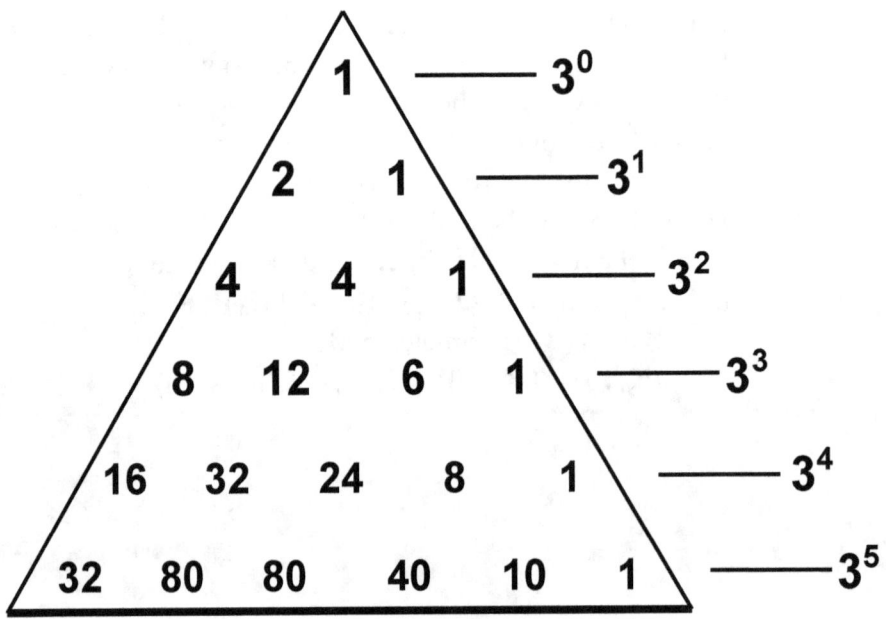

**Duplex Triangle**

# 6

# Another Relative of Pascal's Triangle

## 6.1 The Triplex Triangle

The entries in the rows of **Pascal's Triangle** are binomial coefficients, represented here by C(d, k), the element in row d and column k of the triangular array. They are coefficients of the terms in the expansion of a binomial, such as (n + 1), when raised to some power. For example, if raised to the power **d** and for **j = 0** to **d**, then

$$\textbf{(n + 1)}^{\textbf{d}} = \textbf{C(d, 0)} \cdot \textbf{n}^{\textbf{(d)}} + \textbf{C(d, 1)} \cdot \textbf{n}^{\textbf{(d - 1)}}$$
$$+ \textbf{C(d, 2)} \cdot \textbf{n}^{\textbf{(d - 2)}}$$
$$+ \textbf{C(d, 3)} \cdot \textbf{n}^{\textbf{(d - 3)}}$$
$$+ \ldots + \textbf{C(d, d - 1)} \cdot \textbf{n}^{\textbf{(d - (d - 1))}}$$
$$+ \textbf{C(d, d)} \cdot \textbf{n}^{\textbf{(d - d)}},$$

or, if d = 1 then

$$\textbf{(1 + 1)}^{\textbf{d}} = \textbf{C(d, 0)} + \textbf{C(d, 1)} + \textbf{C(d, 2)} + \textbf{C(d, 3)} + \ldots + \textbf{C(d, d - 1)} + \textbf{C(d, d)},$$

or, equivalently, for **k = 0** to **d**,

$$\textbf{(1 + 1)}^{\textbf{d}} = \Sigma\textbf{C(d, k)} = \textbf{2}^{\textbf{d}}, \tag{1}$$

The rows of the **Simplex Triangle** are the same as those of Pascal's Triangle, but without the leading coefficient of 1. **C(d, k), k = 0** to **d − 1,** also represents the number of simplex sub-elements of the type **k**-simplex in a **d**-simplex.

Recall a **d-duplex** is a block of d-dimensional space (a regular convex polytope) consisting of $2^d$ verticies with d equal and orthogonal edges coincident at each vertex. Additionally, each **d**-duplex consists of d types of sub-elements, all of which are duplexes of dimensions **0** to **d − 1**, that is, points, edges, faces (squares), solids (cubes), and cells of higher dimensions, up to **(d − 1)**-duplexes, of which there are **2d** of them in the **d**-duplex. If the measure of each edge of the **d**-duplex is n, then the measure of capacity of the duplex equals **n^d** and the types and capacities of the sub-elements are represented by **n^{d - k}**, for k = 0 to d − 1. The quantities of each type of sub-element are the coefficients of the terms in the expansion given in equation (1) of Chapter 2, which is restated here.

$$\textbf{(n + 2)}^{\textbf{d}} - \textbf{n}^{\textbf{d}} = \Sigma\textbf{C(d, k)} \cdot \textbf{2}^{\textbf{k}} \cdot \textbf{n}^{\textbf{(d - k)}}, \text{ for } \textbf{k = 1} \text{ to } \textbf{d, d > 0.} \quad \text{(Chap. 2, Eq. 1)}$$

Since we are interested only in the type and number of each sub-element, and not the capacity, we can simply let n = 1 to get

$$(n + 2)^d - n^d = (1 + 2)^d - 1^d = 3^d - 1$$

Furthermore, we wish to include the d-duplex itself in the expansion, so we consider just the expansion of $(1 + 2)^d$. And, finally, we want to show the terms in the expansion representing the sub-elements in increasing value of their dimension from 0 to d, so now we expand $(2 + 1)^d$ instead of $(1 + 2)^d$. (Note: $C(d, k) = C(d, d - k)$ due to the symmetry of the entries in each row of Pascal's Triangle and so the coefficients of the powers of 2 are consistent with the expansion given in Chapter 2.

$$
\begin{aligned}
(2 + 1)^d = &\ C(d, 0) \cdot 2^d + C(d, 1) \cdot 2^{d-1} \\
&+ C(d, 2) \cdot 2^{d-2} + C(d, 3) \cdot 2^{d-3} \\
&+ \ldots + C(d, d - 1) \cdot 2^{(d - (d - 1))} \\
&+ C(d, d) \cdot 2^{d-d}
\end{aligned}
$$

or, equivalently, for **k = 0** to **d**,

$$(2 + 1)^d = \sum C(d, k) \cdot 2^{d-k} = 3^d. \tag{2}$$

Next we define

$$N(d, k) = C(d, k) \cdot 2^{d-k}, \quad k \le d. \tag{3}$$

Formula (3) for **N(d, k)** gives us the entry in the **Duplex Triangle** in row **d** and column **k**. (**N(d, k)** also represents the number of sub-elements of the type **k**-duplex in a **d**-duplex.)

Consider, now, the expansion

$$(3 + 1)^d = \sum C(d, k) \cdot 3^{d-k} = 4^d. \tag{4}$$

and define the coefficient     $P(d, k) = C(d, k) \cdot 3^{d-k}. \tag{5}$

Formula (5) for **P(d, k)**, gives us the entry in a new triangular array in row **d** and column **k**, **k = 0** to **d**. If **d = 0**, then **P(d, k) = 1**. For the want of an appropriate name for now, we will call this the **Triplex Triangle**. The first few rows are as follows:

```
                         1

                     3       1

                 9       6       1         ⸢─── Column 3
                                          /
            27      27       9       1

       81     108      54      12       1

  243    405    270     90      15       1   ⟵── Row 5
```

**Triplex Triangle**

$$P(5, 3) = C(5, 3) \cdot 3^{5-3} = 10 \cdot 3^2 = 90$$

## Section 6.1

1.  Use formula (5) for the Triplex Triangle to find the entry of the 5ᵗʰ row column 2 of the Triangle. Compare the result to the Triangle shown on the previous page.

2.  Use formula (5) for the Triplex Triangle to find all of the entries of 6ᵗʰ row of the Triangle.

Summary:

Equation (1) leads to Pascal's Triangle and the Simplex Triangle:

$$(1 + 1)^d = \sum C(d, k) = 2^d, \text{ for } k = 0 \text{ to } d$$

Equation (2) leads to the Duplex Triangle:

$$(2 + 1)^d = \sum C(d, k) \cdot 2^{d-k} = 3^d, \text{ for } k = 0 \text{ to } d$$

Equation (4) leads to the Triplex Triangle:

$$(3 + 1)^d = \sum C(d, k) \cdot 3^{d-k} = 4^d, \text{ for } k = 0 \text{ to } d$$

## 6.2 Row Sums of The Triplex Triangle

The left member of equation (4) becomes

$$(3 + 1)^d = 4^d.$$

That is, the sum of the elements in row **d** of the Triplex Triangle is equal to **4ᵈ**.

## Section 6.2

1.  Add the elements of each row of the Triplex Triangle shown below and verify that each sum is a power of 4.

```
                          1          _____

                      3       1          _____

                  9       6       1          _____

              27      27      9       1          _____

          81     108      54     12      1          _____

      243    405    270      90     15      1          _____

  729   1458   1215    540    135     18      1          _____
```

2.    Alternately add and subtract  the elements of each row of the Triplex Triangle shown below and verify that each sum is a power of 2.

$$
\begin{array}{c}
1 \qquad \underline{\hspace{2cm}} \\
3 \;-\; 1 \qquad \underline{\hspace{2cm}} \\
9 \;-\; 6 \;+\; 1 \qquad \underline{\hspace{2cm}} \\
27 \;-\; 27 \;+\; 9 \;-\; 1 \qquad \underline{\hspace{2cm}} \\
81 \;-\; 108 \;+\; 54 \;-\; 12 \;+\; 1 \qquad \underline{\hspace{2cm}} \\
243 \;-\; 405 \;+\; 270 \;-\; 90 \;+\; 15 \;-\; 1 \qquad \underline{\hspace{2cm}}
\end{array}
$$

3.    What power of 2 do you get for the alternating sum of row 5 in Exercise 2?

## 6.3   The Triplex Triangle Rule

We also have a rule for the Triplex Triangle, that is similar to Pascal's Rule for Pascal's Triangle and the Duplex Rule for the Duplex Triangle.  Namely,

$$\boxed{\;P(d-1, k-1) \;+\; 3 \cdot P(d-1, k) \;=\; P(d, k).\;} \qquad (6)$$

**Example.**  Let $d = 5$ and $k = 3$.  Then the left-hand side of (6) becomes

$$
\begin{aligned}
P(4, 2) + \mathbf{3} \cdot P(4, 3) &= C(4, 2) \cdot 3^2 + \mathbf{3} \cdot (C(4, 3) \cdot 3^1) \\
&= 6 \cdot 9 + \mathbf{3} \cdot (4 \cdot 3) \\
&= 54 + \mathbf{3} \cdot 12 \\
&= 54 + 36 = 90
\end{aligned}
$$

and the right-hand side of (6) also equals 90,

$$P(5, 3) = C(5, 3) \cdot 3^2 = 10 \cdot 9 = 90.$$

The positions of these entries, that is 54, 12 and 90,  are highlighted in the array shown below.

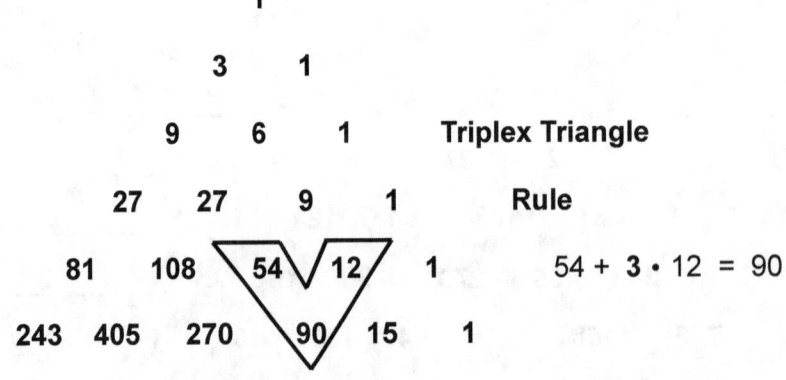

The rule, equation (6), like Pascal's Rule, is a recursive rule.  That is, if we know the entries of row d – 1, then we can find the entries for row d from row d – 1.

**Section 6.3**

1.    Use the recursive rule for the Triplex Triangle to find the entries of $6^{th}$ row of the Triangle, given the entries of the $5^{th}$ row are as shown in the triangular arrays above.

2.    Find the entries of the $7^{th}$ row of the Triplex Triangle.

## 6.4  Extending The Triplex Triangle Rule

The Triplex Rule can be characterized as the <u>string dot product</u> of the <u>reverse</u> of the string formed from row 1 of the Triplex Triangle, and the string of two consecutive entries in another row of the Triplex Triangle, such as, (54, 12), found in the $4^{th}$ row of the Triplex Triangle.  The result is an element in row 5 (row numbers 1 plus 4).

$$(1, 3) \bullet (54, 12) = 54 + 3 \bullet 12 = 54 + 36 = 90 = P(5, 3)$$

For the string dot products of different rows of the Triplex Triangle, we will assume that each row consists of its normal elements and as many additional zeros as necessary to ensure that both strings have an equal number of elements.  However, in the case of the Triplex Triangle,  we will usually reverse the string of the smaller numbered row, when forming string dot products.  Thus, to form the string dot product of the $2^{nd}$ and $4^{th}$ rows of the Triplex Triangle,  we write row 2  with  appended  zeroes and then reverse the row as follows:

$$(0, 0, 1, 6, 9) \bullet (81, 108, 54, 12, 1) = 0 \bullet 81 + 0 \bullet 108 + 1 \bullet 54 + 6 \bullet 12 + 9 \bullet 1$$
$$= 0 + 0 + 54 + 72 + 9$$
$$= 135$$

The result of this string dot product is 135, which is the element in the $6^{th}$ row (2 + 4 = 6) and $4^{th}$ column of the Triplex Triangle. P(6, 4) = 135.

```
                        1

                    3       1

Reverse this row → ( 9    6    1    0    0 )

                   27   27    9    1

                  ( 81  108   54   12    1 )

              243  405  270   90   15    1

          729  1458 1215 540  (135)  18    1           P(6,4) = 135
```

$$(0, 0, 1, 6, 9) \bullet (81, 108, 54, 12, 1) = 0 + 0 + 54 + 72 + 9 = 135$$

In general,

*if we form the string dot product of the n-th and m-th rows of the Triplex Triangle, as described above, we get the element in the (n + m)-th row and the m-th column of the Triplex Triangle.*

### Section 6.4

1.   Compute this string dot product:   $(1, 9, 27, 27) \bullet (27, 27, 9, 1) = ?$
     In what row and column can the result be found in the Triplex Triangle?

2.   Compute string dot product of row 2 and row 3 (reverse the order of the elements in the first string).   In what row and column can the result be found in the Triplex Triangle?

## 6.5   Some Number Patterns Related to the Triplex Triangle

| d | | | P(d, k) | | | | Total | Alt. Sum |
|---|---|---|---|---|---|---|---|---|
| 0 | 1 | | | | | | $1 = 4^0$ | 1 |
| 1 | 3 | 1 | | | | | $4 = 4^1$ | 2 |
| 2 | 9 | 6 | 1 | | | | $16 = 4^2$ | 4 |
| 3 | 27 | 27 | 9 | 1 | | | $64 = 4^3$ | 8 |
| 4 | 81 | 108 | 54 | 12 | 1 | | $256 = 4^4$ | 16 |
| 5 | 243 | 405 | 270 | 90 | 15 | 1 | $1024 = 4^5$ | 32 |
| 6 | 729 | 1458 | 1215 | 540 | 135 | 18 | 1 | $4096 = 4^6$ | 64 |

Column 0 ⟶

**The Triplex Triangle, P(d, k)**

In the Triplex Triangle shown above, observing column 0, we see the pattern

$$P(d, 0) = 3^d$$

and by examining the entries just before the final 1 in each row, we recognize multiples of 3,

$$P(d, d - 1) = 3d.$$

The reader may notice that many of the properties of Duplex Triangle have their analogs here when considering properties of the Triplex Triangle.

Another property, referring to the Triplex Triangle above, relates to even numbered rows.  The element in the middle of an even numbered row is 6 times the element in preceding row and just above the given middle element.  For example, in row 6 the middle element is 540.  The element in the preceding row and just above 540 is 90 and $6 \bullet 90 = 540$. In general,

$$P(2x, x) = 6 \bullet P(2x - 1, x), \quad \text{for } x = 1, 2, 3, \ldots$$

The next property relates to the odd numbered rows. Consider the two elements nearest the middle of the row. One-third of the element on the left equals the element on the right. For example, in row 5, the two elements nearest the middle of the row are **P(5, 2)** = 270 and **P(5, 3)** = 90 and (1/3)· **P(5, 2)** = **P(5, 3)**. In general,

$$(1/3) \cdot P(2x + 1, x) = P(2x + 1, x + 1).$$

In the exercises below we will discover some other interesting properties of the elements in the Triplex Triangle.

---
### Section 6.5

1. Verify the formula **P(2x, x) = 6 · P(2x – 1, x)**,  for $x = 1$ and $x = 2$.
2. Verify the formula **(1/3) · P(2x + 1, x) = P(2x + 1, x + 1)** for $x = 1$ and $x = 3$.
3. Verify the formula **x · P(x – 1, 0) = P(x, 1)**,  for $x = 2, 3, 4, 5$, and 6.
4. Verify the formula **x · 3$^{x-1}$ = P(x, 1)**, for $x = 1, 2, 3$ and 4.
5. Verify the formula **3$^x$ · (x + 1) = P(x + 1, 1)**, for $x = 1, 2, 3$ and 4.
6. Verify the formula **3$^{x-1}$ · P(x + 1, x) = P(x + 1, 1)**, for $x = 1, 2, 3$ and 4.
7. Verify the formula **3$^{x-2}$ · P(x + 2, x) = P(x + 2, 2)**, for $x = 1, 2, 3$ and 4.
8. Verify the formula **3$^{x-3}$ · P(x + 3, x) = P(x + 3, 3)**, for $x = 1, 2, 3$ and 4.
9. Line up the formulas from Exercises 6, 7, and 8. Predict the next formula for that sequence.
10. Choose any row of the Triplex Triangle, for example the 5$^{th}$ row. Separate the elements into two rows as shown.

    ```
                                    405   90    1
    243  405  270  90  15  1  ⟹   243   270   15
    ```

    Form a fraction with the sum of the numbers in the top row as the numerator and the sum of the numbers in the bottom row as the denominator. Add $2^5 = 32$ to the numerator.

    $$\frac{405 + 90 + 1 + \mathbf{32}}{243 + 270 + 15}$$

    The value of this fraction is 1.
    Follow the process described above, adding 2 raised to the power of the row number, for rows 1 through 6 of the Triplex Triangle. What are the values of these fractions?

---

Like Pascal's Triangle and the Duplex Triangle, there are many number patterns in the Triplex Triangle. We have explored a few here and we are confident that there are many more. In the next two sections we will explore additional patterns involving the columns and cross-columns of the Triplex Triangle.

## 6.6  Number Patterns Related to the Cross-columns of the Triplex Triangle

The cross-columns of the Triplex Triangle are indicated below.

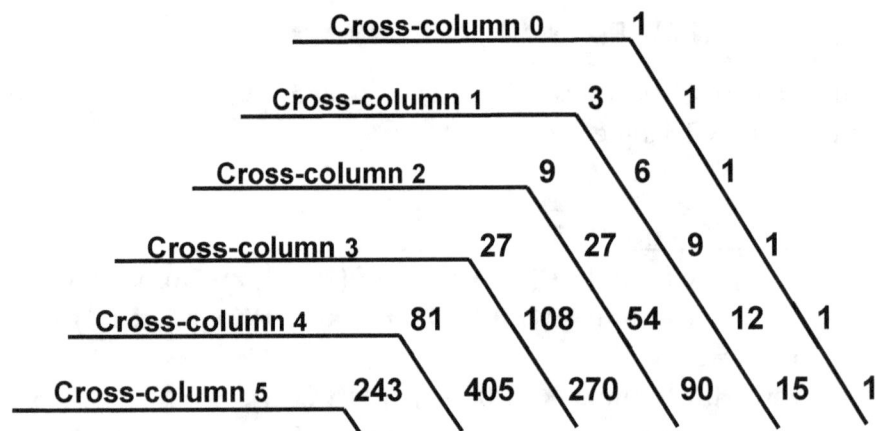

Consider the sums of successive cross-column elements.  For example, the cross-column 2 begins as follows:

9

27

54

90

Now form a sequence of partial sums in the following way:

$$9 = 9 = (1/3) \cdot 27$$
$$9 + 27 = 36 = (1/3) \cdot 108$$
$$9 + 27 + 54 = 90 = (1/3) \cdot 270$$
$$9 + 27 + 54 + 90 = 180 = (1/3) \cdot 540$$

Observe the pattern of partial sums:  9, 36, 90, 180.  If we triple these numbers, we get the beginning of the elements in cross-column 3, that is, 27, 108, 270, 540.  We can write this sum as follows:

**P(2, 0) + P(3, 1) + P(4, 2) + P(5, 3)  =  (1/3) · N(6, 3)**

In general, the sum of the first n + 1 elements of cross-column 2 is equal to one-third the (n + 1)-th element in cross-column 3.  That is,

**P(2, 0) + P(3, 1) + P(4, 2) + P(5, 3) + . . . +  P(n + 2, n)  =  (1/3) · P(n + 3, n)**        (7)

T his is illustrated in the Triplex Triangle shown on the next page for n = 4.

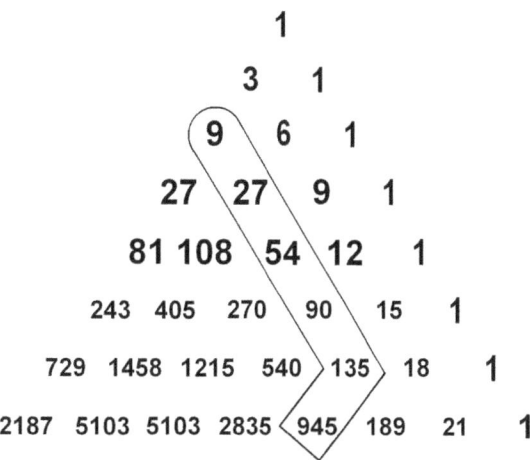

**9 + 27 + 54 + 90 + 135 = 315 = (1/3) • 945**
A property of the cross-columns of the Triplex Triangle,
that is similar to the "hockey-stick" property of Pascal's Triangle.

 ## Section 6.6

1.    **(a)**   What is the sum of the first 3 elements of cross-column 2?
      **(b)**   What is the sum of the first 4 elements of cross-column 2?

2.    Let n = 5 in equation (7). Find the sum of the first 6 elements of cross-column 2. If you triple the sum, in what row and column do you find this value?

3.    Encircle the first 5 elements of cross-column 1 and find their sum. If you triple the sum, in what row and column do you find this value?

4.    Encircle the first 3 elements of cross-column 4 and find their sum. If you triple the sum, in what row and column do you find this value?

5.    Check other cross-columns for the sum of the first n elements. Each time triple the sum and look for the location of that value in the Triplex Triangle.

In general, the sum of the first n + 1 elements of cross-column c of the Triplex Triangle is equal to one-third of the (n + 1)-th element in cross-column c + 1. That is,

$$P(c, 0) + P(c + 1, 1) + P(c + 2, 2) + P(c + 3, 3) + \ldots + P(c + n, n) = (1/3) \cdot P(c + n + 1, n) \quad (8)$$

6.    Let c = 3 and n = 4 in equation (8) above. Find the sum of the first 5 elements of cross-column c. If you triple the sum, in what row and column do you find this value?

7.    Let c = 1 and n = 5 in equation (8) above. Find the sum of the first 6 elements of cross-column c. In what row and column do you find this value? Sketch the corresponding "hockey-stick" on the triangle shown above.

8.    In each of the Triplex Triangles that follow, add all the numbers in the top group, then add all the numbers enclosed in the bottom group. Show in each case that sum of the numbers in the top group is equal to one-third the sum of the numbers in the bottom group.

9.    Add all the numbers in rows 0 through 4 of the Triplex Triangle. Next add all the numbers in row 5 and subtract 1. Show that sum of all of the numbers in rows 0 through 4 is equal to one-third the sum of the numbers in row 5 minus 1.

10.   In the Triplex Triangle shown below, add the numbers in each "hockey stick" handle ( that is, cross-column elements above the gray line) and show that the sum is equal to one-third of the number on that "hockey stick" below the gray line.

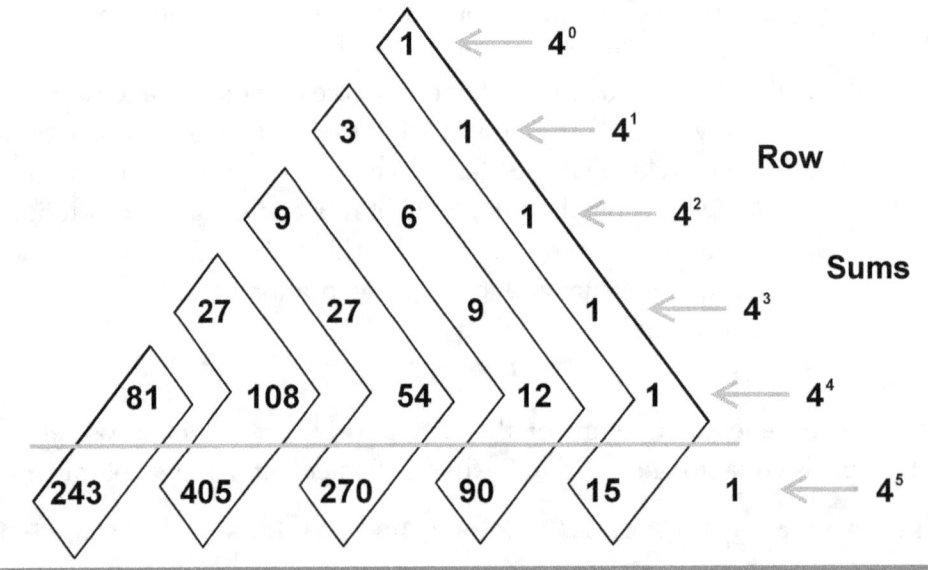

   The sum of all of the elements in the Triplex Triangle up through the 4$^{th}$ row is equal to 1/3 the sum of the elements in the 5$^{th}$ row minus 1. Each row sum is equal to a power of 4. Therefore,

$$4^0 + 4^1 + 4^2 + 4^3 + 4^4 = (1/3)(4^5 - 1).$$

In general, we have a formula for the sum of the powers of 4 as follows:

$$4^0 + 4^1 + 4^2 + \ldots + 4^n = (1/3)(4^{n+1} - 1) \qquad (9)$$

This also gives us the sum of all of the elements in the Triplex Triangle from row 0 through row n, that is, **$(1/3)(4^{n+1} - 1)$**.

---

11. Use the formula (9) to find the sum of the powers of 4 from the $0^{th}$ through the $9^{th}$.

12. Devise a plan to find the sum of all the numbers in the Triplex Triangle from row 6 through row 9.

---

Summary:

The sum of all of the elements in the triangular array from row 0 through row n in

**Pascal's Triangle:**     $(1/1)(2^{n+1} - 1) = 2^0 + 2^1 + 2^2 + \ldots + 2^n$,

**Duplex Triangle:**     $(1/2)(3^{n+1} - 1) = 3^0 + 3^1 + 3^2 + \ldots + 3^n$,

**Triplex Triangle:**     $(1/3)(4^{n+1} - 1) = 4^0 + 4^1 + 4^2 + \ldots + 4^n$.

Consider the sequence of partial sums of the powers of 4.

$$1 = 1$$
$$1 + 4 = 5$$
$$1 + 4 + 16 = 21$$
$$1 + 4 + 16 + 64 = 85$$
$$1 + 4 + 16 + 64 + 256 = 341$$
$$1 + 4 + 16 + 64 + 256 + 1024 = 1365$$

The sequence of partial sums above compares favorably to the sequence of Jacobsthal numbers represented by $J_{2n+2}$, that is, the sequence of Jacobsthal numbers with even numbered subscripts which are as follows:

**1, 5, 21, 85, 341, 1365, . . . , $J_{2n+2}$, . . .**     **for n = 0, 1, 2, 3, . . .**

---

13. Find the sum of the powers of 4 from the $0^{th}$ through the $6^{th}$,
$$1 + 4 + 16 + 64 + 256 + 1024 + 4096$$
and compare to $J_{2n+2}$, n = 6.

14. An explicit formula for $J_n$ is $J_n = (2^n - (-1)^n)/3$. Write the formula for $J_{2n+2}$ and show that it is equal to **$(1/3)(4^{n+1} - 1)$**, which is the sum of the $0^{th}$ to the nth power of 4, or equivalently, the sum of all of the elements in the Triplex Triangle from row 0 through row n.

---

## 6.7 Number Patterns Related to the Columns of the Triplex Triangle

If we add all of the elements in column k of the Triplex Triangle down to the nth row, then add 2 times the sum of all the elements in column k + 1, also down to the nth row, the total is equal to the element in the (n + 1)th row and the column equal to the column number k + 1. For example, the sum of the elements in the 2^nd and 3^rd columns down to the 5^th row is equal to the element in the 6^th row and 3^rd column (see below).

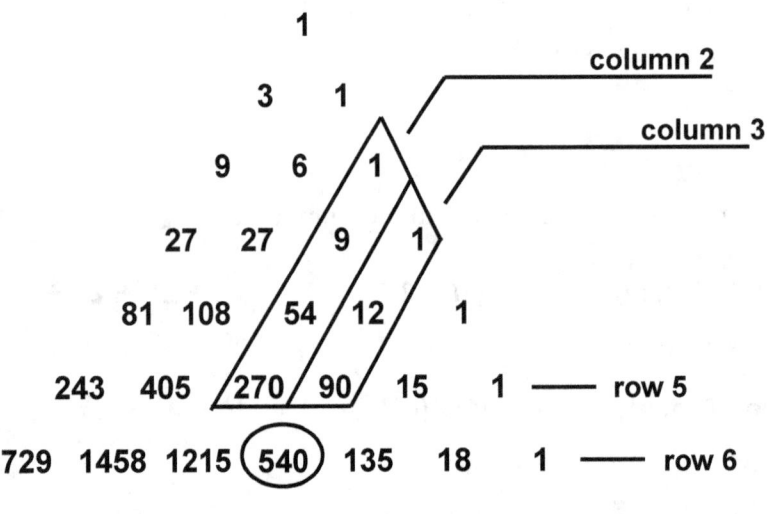

$$(1 + 9 + 54 + 270) + 2 \cdot (1 + 12 + 90) = 540$$

 **Section 6.7**

1. Verify that the sum of the elements in the 1^st and 2 times the 2^nd column down to the 4^th row is equal to the element in the 5^th row and 2^nd column.

2. Verify that the sum of the elements in the 3^rd and 2 times the 4^th column down to the 5^th row is equal to the element in the 6^th row and 4^th column.

3. Experiment with other columns and rows and verify the results.

## 6.8 Number Patterns of the Triplex Triangle Involving String Products and String Dot Products

The **scalar product** of 9 and the 2^nd column of Pascal's Triangle results in the second column of the Triplex Triangle. For example, for the first 4 elements of the second column of both triangles, we have

$$9 (1, 3, 6, 10) = (9, 27, 54, 90).$$

See the figure below:

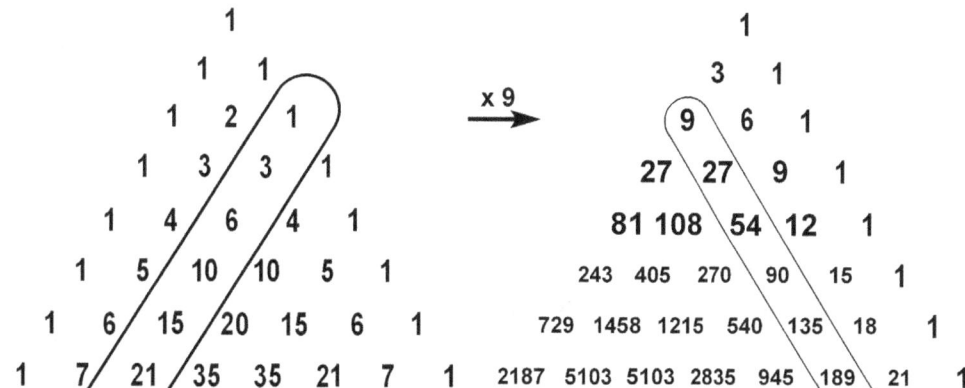

Left: Pascal's Triangle, Column 2 (triangular numbers).
Right: Triplex Triangle, Cross-column 2.
9 times a given triangular number is equal to a corresponding element in the
Triplex Triangle, Cross-column 2.

 **Section 6.8**

1.  Find the scalar product of 9 and the first 5 elements of column 2 of Pascal's
    Triangle, that is, the first 5 triangular numbers. Compare to the first 5
    elements of the cross-column 2 of the Triplex Triangle.
2.  Find the scalar product of 3 and the first 5 elements of column 1 of Pascal's
    Triangle, that is, the first 5 natural numbers. Compare to the first 5
    elements of the cross-column 1 of the Triplex Triangle.
3.  Find the scalar product of 27 and the first 5 elements of column 3 of Pascal's
    Triangle, that is, the first 5 tetrahedral numbers. Compare to the first 5
    elements of the cross-column 3 of the Triplex Triangle.

The 3 exercises above are specific examples of a more general property, namely,

> **the elements of the cross-column c of the Triplex Triangle are equal
> to the scalar product of $3^c$ and the elements of column c of Pascal's
> Triangle.**

4.  Find the scalar product of $3^4$ and the first few elements of column 4 of
    Pascal's Triangle. Compare to the first few elements of the cross-column
    4 of the Triplex Triangle.

In Sec. 6.4 we discussed the extensions of the Triplex Triangle Rule. In the next set of exercises we will see that string dot products can also be formed from strings made up of elements representing only partial rows. For example, form the string dot product of the last 3 three elements of row 3 of the Triplex Triangle with the last three elements of the $4^{th}$ row (we reverse the string representing the $3^{rd}$ row).

$$(1, 9, 27) \bullet (54, 12, 1) = 54 + 108 + 27 = 189 = P(7, 5)$$

The result, 189, is found in the $7^{th}$ row (3 + 4 = 7) and the $5^{th}$ column, P(7, 5) = 189. The column number (5) of the result is determined by adding the Triplex Triangle column positions of the elements in the last positions of the two operand strings. In our example those elements are 27 in the first string and 1 in the second string and their column positions in the Triplex Triangle are 1 and 4, respectively. Adding these numbers gives 5, the column position of the result, 189.

Another example: form the string dot product of the last two elements of the $2^{nd}$ row of the Triplex Triangle with the last two elements of the $4^{th}$ row.

$$(1, 6) \bullet (12, 1) = 12 + 6 = 18$$

This time the result, 18, is found in the $6^{th}$ row (2 + 4 = 6), and in the $5^{th}$ column, P(6, 5) = 18. This is determined by the element 6 in the first operand string, which is in column position **1** of the Triplex Triangle and the element 1 in the second string which is found in the column position **4** in the $4^{th}$ row of the Triplex Triangle (1 + 4 = 5).

---

5.   Compute this string dot product:   (1, 9, 27) • (27, 9, 1) = ?
      In what row and column can the result be found in the Triplex Triangle?

6.   Compute this string dot product:   (1, 6) • (6, 1) = ?
      In what row and column can the result be found in the Triplex Triangle?

7.   Compute this string dot product:   (1, 6, 9) • (54, 12, 1) = ?
      In what row and column can the result be found in the Triplex Triangle?

8.   Compute this string dot product:   (1, 9, 27, 27) • (108, 54, 12, 1) = ?
      In what row and column can the result be found in the Triplex Triangle?

---

Another formula for P(d, k), the element in row d column k of the Triplex Triangle is in the form of a string dot product. The strings in this product come from both Pascal's Triangle and the Duplex Triangle. We will form the product of row d of the Duplex Triangle and column k of Pascal's Triangle.

**P(d, k)  =  (row d) • (column k),**   row d from the Duplex Triangle and column k
from Pascal's Triangle

For example, using row **4** of the Duplex Triangle and column **2** of Pascal's Triangle we have

$$P(4, 2) = (16, 32, 24, 8, 1) \bullet (0, 0, 1, 3, 6)$$
$$= \ 0 + 0 + 24 + 24 + 6$$
$$= \ 54$$

Notice the zeros in the second string. This is because the last element of the second string is required to be in the 4th row, the same row number as the first string used in the product. See the Figure below.

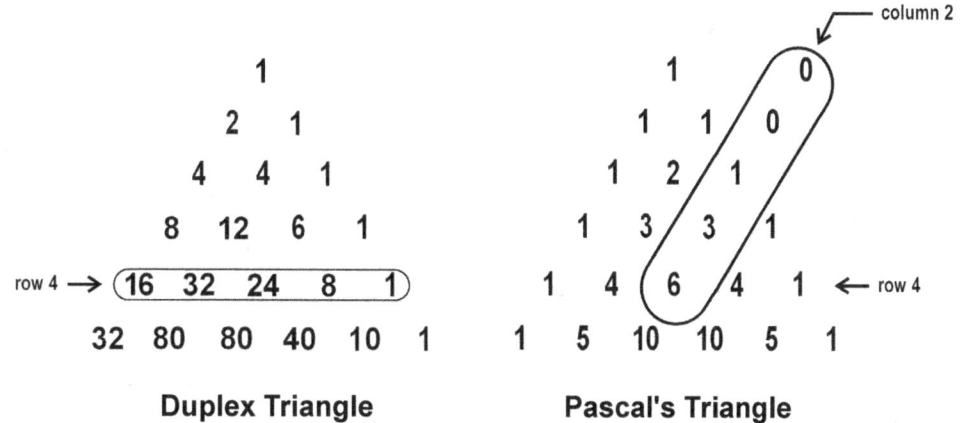

**Duplex Triangle**          **Pascal's Triangle**

The string dot product of a row of the Duplex Triangle and a column **of** Pascal's Triangle equals an element in the Triplex Triangle.

$$P(4, 2) = (16, 32, 24, 8, 1) \bullet (0, 0, 1, 3, 6) = \mathbf{54}$$

---

9.  Compute this string dot product:   $(8, 12, 6, 1) \bullet (0, 1, 2, 3) = ?$
    In what row and column can the result be found in the Triplex Triangle?
10. Compute this string dot product:   $(32, 80, 80, 40, 10, 1) \bullet (0, 0, 1, 3, 6, 10) = ?$
    In what row and column can the result be found in the Triplex Triangle?
11. Compute this string dot product:   The 4th row of the Duplex Triangle and the first column of Pascal's Triangle.
    In what row and column can the result be found in the Triplex Triangle?

---

### Vandermonde's Identity for the Triplex Triangle

In Chapter 3 we demonstrated Vandermonde's Identity for both Pascal's Triangle and the Duplex Triangle. Remarkably, there is a similar property that holds for the Triplex Triangle.

---

**Vandermonde's Identity for the Triplex Triangle**

Choose row r of the Triangle. Split r into two parts, m and n, so that m + n = r. Next select an element anywhere in that row, say in column position c. Then P(r, c) is equal to the string dot product of the string of elements in row m from column 0 to column c and the string of elements in row n from column c back to column 0.

---

For example, choose row 6 of the Triplex Triangle, $r = 6$, and split 6 into two parts. We will arbitrarily split it into the parts 4 and 2, so $m = 4$ and $n = 2$ and $m + n = r = 6$. Now select a column position in row 6, we will arbitrarily choose $c = 3$. We now have the element $P(r, c) = P(6, 3) = C(6, 6 - 3) \cdot 3^{6-3} = 20 \cdot 27 = 540$. Vandermonde's identity states that this element is equal to the string dot product of the string in the m th row ($4^{th}$ row) of the Triplex Triangle from column 0 to $c = 3$, and the string in the nth row ($2^{nd}$ row) from column $c = 3$ down to 0. That is,

$$P(6, 3) = (81, 108, 54, 12) \cdot (0, 1, 6, 9)$$
$$= 81 \cdot 0 + 108 \cdot 1 + 54 \cdot 6 + 12 \cdot 9$$
$$= \quad 0 \quad + \quad 108 \quad + \quad 324 \quad + \quad 108$$
$$= \quad 540$$

Note: If the smaller row of the Triplex Triangle does not have a enough column elements to match the longer row, then 0's are **prepended** into the string of elements corresponding to the shorter row to make the number of positions match the longer string.

12.   Select row 6 of the Triplex Triangle, $r = 6$, and split it into two parts so $m = 3$ and $n = 3$ and $m + n = r = 6$. Let the column position in row 6 equal 3, $c = 3$. Form the two strings described in Vandermonde's Identity for the Triplex Triangle and show that the string dot product equals $P(6, 3) = 540$.

13.   Select row 6 of the Triplex Triangle, $r = 6$, and split it into two parts so $m = 5$ and $n = 1$ and $m + n = r = 6$. Let the column position in row 6 equal 3, $c = 3$. Form the two strings described in Vandermonde's Identity for the Triplex Triangle and show that the string dot product equals $P(6, 3) = 540$.

14.   Select row 6 of the Triplex Triangle, $r = 6$, and split it into two parts so $m = 3$ and $n = 3$ and $m + n = r = 6$. Let the column position in row 6 equal 5, $c = 5$. Form the two strings described in Vandermonde's Identity for the Triplex Triangle and show that the string dot product equals $P(6, 5) = 18$.

## 6.9   Diagonal Patterns Within the Triplex Triangle

The sums of the elements on each of the diagonals of the Triplex Triangle form a sequence of numbers, and like the Fibonacci sequence, a formula for this new sequence is a recursive formula where each element is based on the previous two elements in the sequence. The sequence is

**1, 3, 10, 33, 109, 360, 1189, . . ., $R_n$ , . . .**

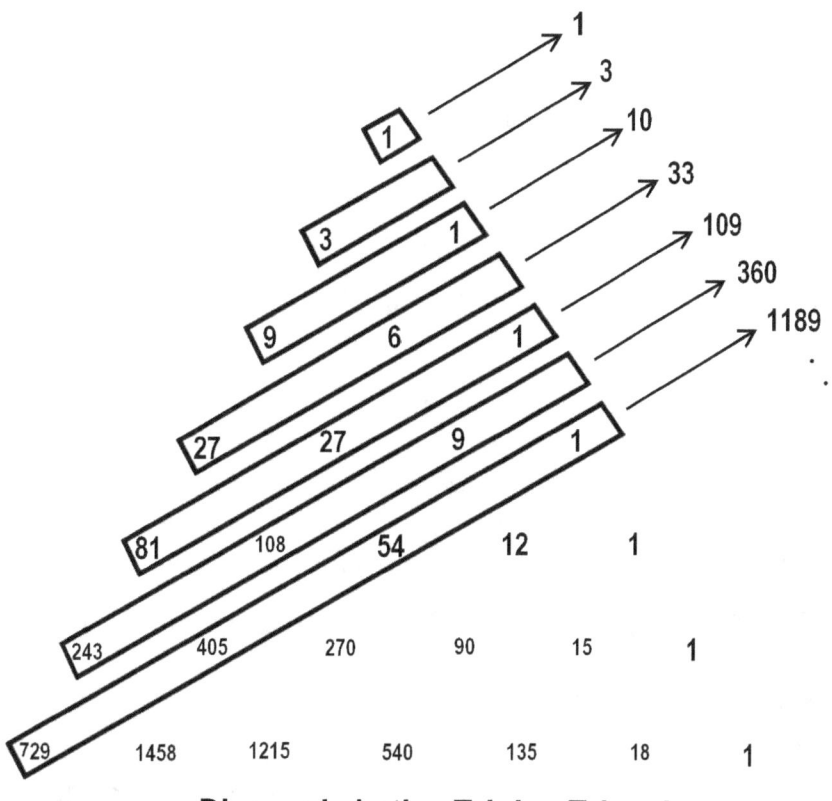

**Diagonals in the Triplex Triangle**

Unlike the Fibonacci and Pell sequences, an element in this sequence is equal to **3** times the previous element plus the element just prior to that. The recursive rule is

$$R_n = 3 \cdot R_{n-1} + R_{n-2}$$

for n = 2, 3, . . . and $R_0 = 0$ and $R_1 = 1$.

For example, $R_5 = 3 \cdot R_4 + R_3$, or equivalently, 109 = 3·33 + 10.

*The sequence $R_n$ is to the Triplex Triangle*
*as the Fibonacci and Pell sequences are to Pascal's Triangle*
*and the Duplex Triangle, respectively.*

The recursive rule is based on row 1 of the Triplex Triangle, in that it is the <u>string dot product</u> of that row with two previous consecutive elements in the sequence. This gives us the following alternative version of the recursive rule:

$$R_n = (3, 1) \bullet (R_{n-1}, R_{n-2}) = 3 \cdot R_{n-1} + 1 \cdot R_{n-2}.$$

In general, we have the following recursive formulas, which are the Triplex extensions of this recursive rule.

$$R_n = \; 1 \cdot R_n$$

$$= \; 3 \cdot R_{n-1} + \; 1 \cdot R_{n-2} \tag{10}$$

$$= \; 9 \cdot R_{n-2} + \; 6 \cdot R_{n-3} + \; 1 \cdot R_{n-4} \tag{11}$$

$$= \; 27 \cdot R_{n-3} + \; 27 \cdot R_{n-4} + \; 9 \cdot R_{n-5} + \; 1 \cdot R_{n-6} \tag{12}$$

$$= \; 81 \cdot R_{n-4} + 108 \cdot R_{n-5} + \; 54 \cdot R_{n-6} + 12 \cdot R_{n-7} + \; 1 \cdot R_{n-8} \tag{13}$$

$$= 243 \cdot R_{n-5} + 405 \cdot R_{n-6} + 270 \cdot R_{n-7} + 90 \cdot R_{n-8} + 15 \cdot R_{n-9} + 1 \cdot R_{n-10} \tag{14}$$

etc.

Formulas (10) thru (14) represent the <u>string dot products</u> of the rows of the Triplex Triangle with consecutive elements of the $R_n$ sequence. For instance, if we use row 3 of the Triplex Triangle in the product, then the consecutive sequence of $R_n$ numbers begins with $R_{n-3}$ and works backward for three more of the sequence numbers, or a total of four consecutive sequence numbers (in this case it is assumed that n is at least 6):

$$R_n = (27, 27, 9, 1) \bullet (R_{n-3}, R_{n-4}, R_{n-5}, R_{n-6}) = 27 \cdot R_{n-3} + 27 \cdot R_{n-4} + 9 \cdot R_{n-5} + 1 \cdot R_{n-6}.$$

**Example:** Verify formula (11) for $n = 6$.
**Solution:** $R_6 = 360$ and

$$9 \cdot R_{n-2} + 6 \cdot R_{n-3} + 1 \cdot R_{n-4} = 9 \cdot R_4 + 6 \cdot R_3 + 1 \cdot R_2$$
$$= 9 \cdot 33 + 6 \cdot 10 + 1 \cdot 3$$
$$= 297 + 60 + 3$$
$$= 360$$

We can also prove formula (11) by using formula (10), the given recursive rule for this sequence.

Given Formula (10):  $R_n = \; 3 \cdot R_{n-1} + 1 \cdot R_{n-2}$
Use the recursive rule and replace $R_{n-1}$ and $R_{n-2}$ with
$3 \cdot R_{n-2} + 1 \cdot R_{n-3}$  and  $3 \cdot R_{n-3} + 1 \cdot R_{n-4}$, respectively.

Therefore,

$$R_n = \; 3 \cdot R_{n-1} + 1 \cdot R_{n-2}$$
$$= \; 3 \cdot [3 \cdot R_{n-2} + 1 \cdot R_{n-3}] + 1 \cdot [3 \cdot R_{n-3} + 1 \cdot R_{n-4}]$$
$$= \; 9 \cdot R_{n-2} + 3 \cdot R_{n-3} + 3 \cdot R_{n-3} + 1 \cdot R_{n-4}$$
$$= \; 9 \cdot R_{n-2} + 6 \cdot R_{n-3} + 1 \cdot R_{n-4} \qquad \blacksquare$$

**Section 6.9**

1.   Verify formula (11) for **(a)** $n = 4$ and **(b)** $n = 5$.
2.   Verify formula (12) for **(a)** $n = 7$ and **(b)** $n = 8$.

3.  Prove formula (12) by using formula (10), the recursive rule for the sequence.

4.  Verify formula (13) for $n = 8$.

5.  Write the string dot product using row 6 of the Triplex Triangle and the the appropriate sequence of consecutive numbers of the $R_n$ sequence.

The recursive formula $R_n = 3 \cdot R_{n-1} + R_{n-2}$ also fits nicely into another pattern of recursive forumlas for $R_n$.

$$R_n = \quad 3 \cdot R_{n-1} + \quad 1 \cdot R_{n-2} = R_2 \cdot R_{n-1} + R_1 \cdot R_{n-2}, \; n \geq 2 \tag{15}$$

$$= \quad 10 \cdot R_{n-2} + \quad 3 \cdot R_{n-3} = R_3 \cdot R_{n-2} + R_2 \cdot R_{n-3}, \; n \geq 3 \tag{16}$$

$$= \quad 33 \cdot R_{n-3} + \quad 10 \cdot R_{n-4} = R_4 \cdot R_{n-3} + R_3 \cdot R_{n-4}, \; n \geq 4 \tag{17}$$

$$= 109 \cdot R_{n-4} + \quad 33 \cdot R_{n-5} = R_5 \cdot R_{n-4} + R_4 \cdot R_{n-5}, \; n \geq 5 \tag{18}$$

$$= 360 \cdot R_{n-5} + 109 \cdot R_{n-6} = R_6 \cdot R_{n-5} + R_5 \cdot R_{n-6}, \; n \geq 6 \tag{19}$$

etc.

**Example:** Verify formula (17) for $n = 6$.
**Solution:** $R_6 = 360$ and

$$
\begin{aligned}
R_4 \cdot R_{n-3} + R_3 \cdot R_{n-4} &= 33 \cdot R_{n-3} + 10 \cdot R_{n-4} \\
&= 33 \cdot R_3 \quad + 10 \cdot R_2 \\
&= 33 \cdot 10 \quad + 10 \cdot 3 \\
&= 330 \quad\quad + 30 \\
&= 360
\end{aligned}
$$

6.  Verify formula (17) for $n = 5$.
7.  Verify formula (18) for $n = 7$.
8.  Predict the next formula to follow formula (19).

**In general, for given n and j, $j \leq n$,**

$$R_n = R_j \cdot R_{n-(j-1)} + R_{j-1} \cdot R_{n-j}, \; n \geq j \tag{20}$$

9.  Verify formula (20) for $n = 8$ and $j = 7$.

Let $j = n + 1$ in formula (20), then a formula similar to formula (20) is

$$R_{2n} = R_{n+1} \cdot R_n + R_n \cdot R_{n-1} = R_n \cdot (R_{n+1} + R_{n-1}),$$

or equivalently,

$$\frac{R_{2n}}{R_n} = R_{n+1} + R_{n-1}. \qquad (21)$$

---

**10.** Verify formula (21) for $n = 2, 3,$ and 4.

---

Consider the **rth** row of the Triplex Triangle (which consists of $r + 1$ elements), then choose $r + 1$ consecutive numbers from the $R_n$ sequence derived from the diagonals of Triplex Triangle (diagonal sequence). Now form the **string dot product** of the <u>reverse</u> of the **rth** row and the string of $r + 1$ consecutive numbers from the $R_n$ sequence. If the last number in the string equals **p** and is in position n of the $R_n$ sequence, then the result of the product equals $R_{r+n}$.

**Example:** Choose row 2 of the Triplex Triangle and $2 + 1 = 3$ consecutive numbers from the $R_n$ sequence starting with 10, that is, $(10, 33, 109)$. Then $p = 109$ and is, therefore, in position 5. So, $R_{2+5} = R_7 = 1189$. Now reverse the 2nd row of the Triplex Triangle to get $(1, 6, 9)$. The result of the string dot product is also 1189.

$$(1, 6, 9) \bullet (10, 33, 109) = 1 \cdot 10 + 6 \cdot 33 + 9 \cdot 109$$
$$= 10 + 198 + 981$$
$$= 1189$$

---

**11.** Form the string of the reverse of the elements in row 2 of the Triplex Triangle and the string $(3, 10, 33)$ of 3 consecutive elements of the $R_n$ sequence. Find the string dot product of these two strings and show that it is equal to $R_6 = 360$.

**12.** Choose row 3 of the Triplex Triangle and $3 + 1 = 4$ consecutive numbers of the $R_n$ sequence starting with $R_2 = 3$. Form the string dot product of the reverse of row 3 of the Triplex Triangle and the string of 4 consecutive numbers of the $R_n$ sequence starting with $R_2 = 3$. Verify that the position of last number in the second string is in position $n = 5$ and that the result of the string dot product is equal to $R_{3+5} = R_8 = 3927$.

**13.** Form the string consisting of $R_6$ and $R_7$. Use row 1 of the Triplex Triangle to form the second string (in reverse) and show that the string dot product of the two strings is equal to $R_8$.

---

There are other properties of the numbers in the $R_n$ sequence that fit into patterns that also work for the Fibonacci and Pell numbers. For example, the formulas

$$(F_n)^2 - (F_{n-1}) \cdot (F_{n+1}) = \pm 1 \quad \text{and} \quad (P_n)^2 - (P_{n-1}) \cdot (P_{n+1}) = \pm 1$$

are true for Fibonacci and Pell numbers as is the corresponding formula for the $R_n$ numbers:

$$(R_n)^2 - (R_{n-1}) \cdot (R_{n+1}) = \pm 1.$$

And the formulas

$$(F_n + 1) \cdot (F_n - 1) = (F_{n-1}) \cdot (F_{n+1}) \quad \text{and} \quad (P_n + 1) \cdot (P_n - 1) = (P_{n-1}) \cdot (P_{n+1})$$

are true if n is an odd number greater than 1, while the corresponding formula for the $R_n$ numbers is

$$(R_n + 1) \cdot (R_n - 1) = (R_{n-1}) \cdot (R_{n+1})$$

and is true if n is an odd number greater than 1.

---

14.    Verify the formula $(R_n)^2 - (R_{n-1}) \cdot (R_{n+1}) = \pm 1$ for n = 3, 4, and 5.
15.    Choose any odd number greater than 1 for n and show that
$$(R_n + 1) \cdot (R_n - 1) = (R_{n-1}) \cdot (R_{n+1}).$$
16.    Choose any even number greater than 0 for n and show that
$$(R_n + 1) \cdot (R_n - 1) - (R_{n-1}) \cdot (R_{n+1}) = 2.$$

---

The formulas

$$(F_n)^2 + (F_{n+1})^2 = F_{2n+1} \quad \text{and} \quad (P_n)^2 + (P_{n+1})^2 = P_{2n+1}$$

are true for Fibonacci and Pell numbers as is the corresponding formula for the numbers $R_n$:

$$(R_n)^2 + (R_{n+1})^2 = R_{2n+1}$$

---

17.    Verify the formula $(R_n)^2 + (R_{n+1})^2 = R_{2n+1}$ for n = 1, 2, 3 and 4.
18.    Verify the formula $(R_n)^2 = 9 \cdot (R_{n-1})^2 + 6 \cdot R_{n-1} \cdot R_{n-2} + 1 \cdot (R_{n-2})^2$
      for n = 3, 4 and 5.

---

The product of two consecutive numbers of the $R_n$ sequence, $R_n \cdot R_{n+1}$, is equal to 3 times the sum of the squares of those numbers from $R_1$ up to $R_n$, that is,

$$R_n \cdot R_{n+1} = 3[(R_1)^2 + (R_2)^2 + (R_3)^2 + \ldots + (R_n)^2] \tag{22}$$

---

19.    Verify formula (22) for n = 3, 4, and 5.

---

Note: There are formulas similar to (22) for the Fibonacci and Pell numbers (Chapter 4),

$$F_n \cdot F_{n+1} = 1[(F_1)^2 + (F_2)^2 + (F_3)^2 + \ldots + (F_n)^2]$$

and

$$P_n \cdot P_{n+1} = 2[(P_1)^2 + (P_2)^2 + (P_3)^2 + \ldots + (P_n)^2].$$

The **string dot product** of the string of Fibonacci numbers from the first up to the nth and the string of $R_n$ numbers from the nth down to the first is equal to one-half the difference between $R_{n+1}$ and the (n + 1)th Fibonacci number. That is,

$$(1, 1, 2, 3, 5, \ldots, F_n) \bullet (R_n, R_{n-1}, R_{n-2}, \ldots, 33, 10, 3, 1)$$

$$= 1 \cdot R_n + 1 \cdot R_{n-1} + 2 \cdot R_{n-2} + 3 \cdot R_{n-3} + 5 \cdot R_{n-4} +$$
$$\ldots + F_{n-3} \cdot 33 + F_{n-2} \cdot 10 + F_{n-1} \cdot 3 + F_n \cdot 1$$
$$= (R_{n+1} - F_{n+1}) / 2. \tag{23}$$

20.    Verify the formula $(1, 1, 2, 3, 5) \bullet (109, 33, 10, 3, 1) = (R_6 - F_6)/2$
21.    Verify the formula (23) for n = 3 and n = 4.

## 6.10   The $R_n$ – Pythagorean Triples

In Sections 4.2 and 5.8 we discussed Pythagorean triples in connection with the Fibonacci, Pell and Jacobsthal sequences of numbers.  The sides of the right triangle, **(a, b, c)**, are determined by **Euclid's formulas:**

$$a = x^2 - y^2, \quad b = 2xy, \quad \text{and} \quad c = x^2 + y^2$$

where $x > y$ and **a** and **b** are the legs of the right triangle and **c** is the hypotenuse, so that

$$c^2 = a^2 + b^2.$$

Now if we choose consecutive $R_n$ numbers for x and y, then c, the hypotenuse, will also be a $R_n$ number.   Examine the table below:

| x | y | $a = x^2 - y^2$ | $b = 2xy$ | $c = x^2 + y^2$ | |
|---|---|---|---|---|---|
| $R_2 = \ \ 3$ | 1 | 8 | 6 | $10 = R_3$ | ⟵ row 1 |
| $R_3 = \ \ 10$ | 3 | 91 | 60 | $109 = R_5$ | |
| $R_4 = \ \ 33$ | 10 | 989 | 660 | $1189 = R_7$ | |
| $R_5 = 109$ | 33 | 10792 | 7194 | $12970 = R_9$ | |
| $\ldots$ | $\ldots$ | | | | |

Notice that the subscripts of the $R_n$ numbers in the last column are odd numbers and are equal to the sum of the subscripts of $R_n$ numbers represented by x and y.  Note that the difference, $c - b$, is the square of numbers in the sequence $1, 2, 7, 23, 76, \ldots, r_n, \ldots$, where $r_n = 3 \cdot r_{n-1} + r_{n-2}$ for $n > 2$ and $r_1 = 1$, $r_2 = 2$.  We will call this sequence the "**first companion $R_n$ sequence**", Also note that the sum, $c + b$, is the square of numbers in the sequence $1, 4, 13, 43, 142, \ldots, t_n, \ldots$, where $t_n = 3 \cdot t_{n-1} + t_{n-2}$ for $n > 2$ and $t_1 = 1$, $t_2 = 4$.  We will call this sequence the "**second companion $R_n$ sequence**".

### Section 6.10

1. Extend the table above by forming the next line in the sequence.
2. Verify that $R_{11}$ is equal to the value of c in your extension of the table.
3. Verify that the difference, **c – b,** in each line of the table involving the $R_n$ numbers, is the square of a number in the sequence called the "first companion $R_n$ sequence", that is, **1, 2, 7, 23, 76, . . . , $r_n$, . . . .**
4. Verify that the sum, **c + b,** in each line of the table involving the $R_n$ numbers, is the square of a number in the sequence called the "second companion $R_n$ sequence", that is, **1, 4, 13, 43, 142, . . . , $t_n$, . . . .**
5. Show that the first few elements of the sequence called the "first companion $R_n$ sequence", that is, **1, 2, 7, 23, 76, . . . , $r_n$, . . . ,** are equal to differences of two consecutive $R_n$ numbers.
6. Show that the first few elements of the sequence called the "second companion $R_n$ sequence", that is, **1, 4, 13, 43, 142, . . . , $t_n$, . . . ,** are equal to sums of two consecutive $R_n$ numbers.
7. For each row of the table involving the $R_n$ numbers, show that **(a)** the perimeter of the triangle, **a + b + c,** for that row is equal to $2 \cdot R_{n+1} \cdot t_{n+1}$, and **(b)** the area is equal to $(R_n \cdot r_{n+1})/2$ times the perimeter.
8. Show that $\mathbf{a} = r_{n+1} \cdot t_{n+1}$, for n = 1, 2, 3, 4 and 5.
9. Show that $\mathbf{b} = 6 \cdot [(R_1)^2 + (R_2)^2 + (R_3)^2 + \ldots + (R_n)^2]$, for n = 1, 2, 3, 4 and 5.
10. Show that $\mathbf{c} = (R_n)^2 + (R_{n+1})^2 = R_{2n+1}$, for n = 1, 2, 3, 4 and 5.

## 6.11   The Sum of the $R_n$ Numbers

The formula for the sum of the $R_n$ numbers fits nicely into a patterned sequence for the sums of the Fibonacci and Pell sequences (Section 4.5).

$$\text{sum of Fibonacci numbers} = (F_{n+1} + F_n - 1)/1,$$

$$\text{sum of Pell numbers} = (P_{n+1} + P_n - 1)/2,$$

$$\text{sum of the } R_n \text{ numbers} = (R_{n+1} + R_n - 1)/3.$$

### Section 6.11

1. Find the sum **1 + 3 + 10 + 33 + 109.** Then use the formula above for the sum of the $R_n$ numbers, n = 5, and verify the same result.
2. Find the sum **1 + 3 + 10 + 33 + . . . + $R_n$** for n = 6 and 7 using the formula above for the sum of the $R_n$ numbers.

3.    Complete the partial sums of the $R_n$ numbers shown below.

$$1 = 1$$
$$1 + 3 = 4$$
$$1 + 3 + 10 = 14$$
$$1 + 3 + 10 + 33 = 47$$
$$1 + 3 + 10 + 33 + 109 = 156$$

$$1 + 3 + 10 + 33 + 109 + 360 = \underline{\quad}$$

$$1 + 3 + 10 + 33 + 109 + 360 + 1189 = \underline{\quad}$$

The sequence of partial sums of the $R_n$ numbers (started in Exercise 3 above) is also a recursive sequence:

$$1, 4, 14, 47, 156, \ldots, U_n, \ldots$$

where

$$U_n = 3 \cdot U_{n-1} + 1 \cdot U_{n-2} + 1$$

and $U_1 = 1$ and $U_2 = 4$ and $n = 3, 4, 5, \ldots$.

4.    Verify the formula $U_n = 3 \cdot U_{n-1} + 1 \cdot U_{n-2} + 1$ for $n = 3, 4,$ and 5.
5.    Verify the formula $U_n = (R_{n+1} + R_n - 1)/3$ for $n = 3, 4,$ and 5.
6.    Verify the formula $R_n = 2 \cdot U_{n-1} + U_{n-2} + 1$ for $n = 3, 4,$ and 5.

We have a relationship with the Triplex Triangle and the $R_n$ numbers similar to the relationships we explored in Chapter 4 with Pascal's Triangle and the Fibonacci diagonals and the Duplex Triangle and the Pell diagonals.

We first look at the sum of the $R_n$ numbers from $R_1$ to $R_n$ equal to $(R_{n+1} + R_n - 1)/3$. Also, the sum of all the elements in the Triplex Triangle from row 0 to row $n - 1$ is equal to $(1/3)(4^n - 1)$ (formula (9)).

Since the sum of the $R_n$ numbers, as found on the $R_n$ diagonals in the Triplex Triangle correspond to elements enclosed by the triangle shown, the formula for the sum of the Triplex Triangle elements **not** represented by $R_n$ numbers for a given value of n, is equal to the sum of all the elements of the Triplex Triangle up to row $(n - 1)$ minus the sum of the $R_n$ numbers from $R_1$ to $R_n$. That is,

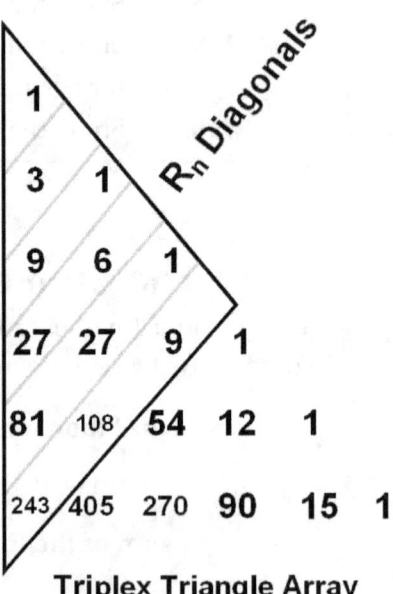

**Triplex Triangle Array**

$$(1/3)(4^n - 1) - (R_{n+1} + R_n - 1)/3.$$

In the figure shown above, we have $n = 6$ and the sum of the $R_n$ numbers from $R_1$ to $R_6$

(those elements inside the drawn triangle) equals

$$1 + 3 + 10 + 33 + 109 + 360 = 516,$$

or, by the formula,

$$(R_{n+1} + R_n - 1)/3 = (R_7 + R_6 - 1)/3 = (1189 + 360 - 1)/3 = 516.$$

The sum of all of the elements in the Triplex Triangle up through the 5$^{th}$ row is equal to

$$(1/3)(4^{r+1} - 1) = (1/3)(4^6 - 1) = (1/3)(4096 - 1) = 1365.$$

The sum of the remaining elements of the Duplex Triangle, those outside the elements enclosed in the $R_n$ diagonals, is equal to $1365 - 516 = 849$.

Therefore, the expression given above for the sum of those elements, that is the formula

$$(1/3)(4^n - 1) - (R_{n+1} + R_n - 1)/3,$$

for $n = 6$ is

$$(1/3)(4^6 - 1) - (R_7 + R_6 - 1)/3 = 1365 - 516 = 849.$$

This is verified by adding all of the elements of the Triplex Triangle shown above that are **not** enclosed by the Pell diagonals:

$$1 + (54 + 12 + 1) + (405 + 270 + 90 + 15 + 1) = 1 + 67 + 781 = 849.$$

---

7.      For each of the figures below find **(a)** the sum of all of the elements shown in the Triplex Triangle, **(b)** the sum of the elements enclosed in the drawn triangle (sum of the elements on the $R_n$ diagonals), and **(c)** the sum of the elements **not** enclosed in the triangle.

In each case, compare your sums to the values of the expressions

        (1)   $(1/3)(4^n - 1)$,          (2)   $(R_{n+1} + R_n - 1)/3$,

and    (3)   $(1/3)(4^n - 1) - (R_{n+1} + R_n - 1)/3$,

where n equals one more than the row number of the last row shown in each figure.

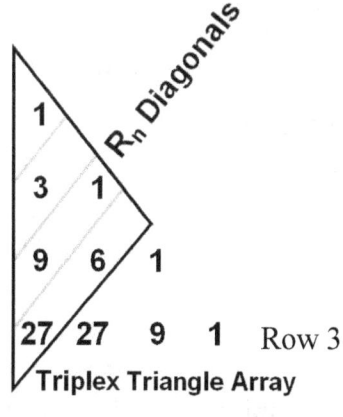

Row 3

Triplex Triangle Array

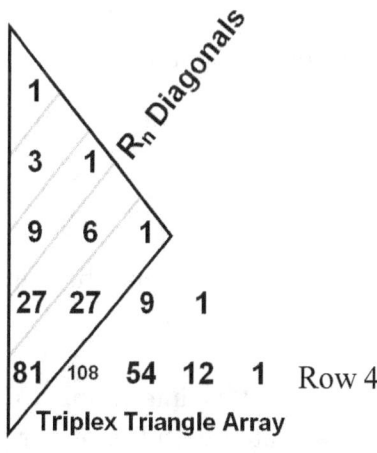

Row 4

Triplex Triangle Array

The expression $(1/3)(4^n - 1) - (R_{n+1} + R_n - 1)/3$, given above for the sum of those elements of the Triplex Triangle that are **not enclosed** with the $R_n$ diagonals, provides a formula for a sequence of numbers which we will denote by

$$e(n) = (1/3)(4^n - 1) - (R_{n+1} + R_n - 1)/3, \text{ for } n = 1, 2, 3, \ldots.$$

This produces the sequence

$$0, 1, 7, 38, 185, 849, 3756, \ldots, e(n), \ldots.$$

Also, notice that $e(4) = 38 = 4 \cdot e(3) + 10$ and $e(5) = 185 = 4 \cdot e(4) + 33$. The 10 and 33 that were added in those two formulas are the $R_n$ numbers, $R_3$ and $R_4$. Noting this we can write $e(4) = 4 \cdot e(3) + R_3$ and $e(5) = 4 \cdot e(4) + R_4$.

---

8.   Complete the following table. Compare the final values of columns 3 and 4 with the values in column 2. Make a conjecture.

| $n$ | $e(n)$ | $(1/3)(4^n - 1) - (R_{n+1} + R_n - 1)/3$ | $4 \cdot e(n-1) + R_{n-1}$ |
|---|---|---|---|
| 1 | 0 | $(1/3)(4^1 - 1) - (R_2 + R_1 - 1)/3 = 0$ | $4 \cdot e(0) + R_0 = 0$ |
| 2 | 1 | $(1/3)(4^2 - 1) - (R_3 + R_2 - 1)/3 = 1$ | $4 \cdot e(1) + R_1 = 1$ |
| 3 | 7 | $(1/3)(4^3 - 1) - (R_4 + R_3 - 1)/3 = 7$ | $4 \cdot e(2) + R_2 = 7$ |
| 4 | 38 | $(1/3)(4^4 - 1) - (R_5 + R_4 - 1)/3 = 38$ | $4 \cdot e(3) + R_3 = 38$ |
| 5 | 185 | $(1/3)(4^5 - 1) - (R_6 + R_5 - 1)/3 = \underline{\quad}$ | $4 \cdot e(4) + R_4 = \underline{\quad}$ |
| 6 | 849 | $(1/3)(4^6 - 1) - (R_7 + R_6 - 1)/3 = \underline{\quad}$ | $4 \cdot e(5) + R_5 = \underline{\quad}$ |
| 7 | 3756 | $\underline{\hspace{3cm}} = \underline{\quad}$ | $\underline{\hspace{2cm}} = \underline{\quad}$ |
| 8 | $\underline{\quad}$ | $\underline{\hspace{3cm}} = \underline{\quad}$ | $\underline{\hspace{2cm}} = \underline{\quad}$ |

---

In general,

$$e(n) = (1/3)(4^n - 1) - (R_{n+1} + R_n - 1)/3 = 4 \cdot e(n-1) + R_{n-1}. \quad (24)$$

Another formula for the general term of the sequence $e(n)$ is

$$e(n) = e(n-1) + 4^{n-1} - R_n. \quad (25)$$

---

9.   Evaluate formula (25) for $n = 3, 4$, and 5. Verify the values found with those in the table above.

---

It is interesting to note the resemblance of the formula for $a(n)$, the sequence associated with Pascal's Triangle and the Fibonacci diagonals and the formula for $b(n)$, the sequence associated with the Duplex Triangle and the Pell diagonals (Chapter 4) and the formula for **c(n)**, the sequence associated with the Duplex Triangle and the Jacobsthal anti-diagonals (Chapter 5). Compare with the formula for **e(n)**.

$$a(n) = 2 \cdot a(n-1) + F_{n-1},$$

anti-diagonals

$$b(n) = 3 \cdot b(n-1) + P_{n-1}, \quad c(n) = 3 \cdot c(n-1) + 2 \cdot J_{n-1},$$

$$e(n) = 4 \cdot e(n-1) + R_{n-1}.$$

## 6.12 Anti-Diagonal Patterns Within the Triplex Triangle

The sums of the elements on each of the anti-diagonals of the Triplex Triangle form a sequence of numbers, and like the Jacobsthal sequence, a formula for this new sequence is a recursive formula where each element is based on the previous two elements in the sequence. The sequence is

**1, 1, 4, 7, 19, 40, 97, . . ., S$_n$ , . . .**

Unlike the Jacobsthal sequence, an element in this sequence is equal to the previous element plus **3** times the element just prior to that. The recursive rule is

$$S_n = S_{n-1} + 3 \cdot S_{n-2}$$

for $n = 2, 3, \ldots$ and **S$_0$ = 0** and **S$_1$ = 1.**

**Anti-Diagonals in the Triplex Triangle**

For example, **S$_5$ = S$_4$ + 3 · S$_3$**, or equivalently, **19 = 7 + 3 · 4.**

*The sequence $S_n$ is to the Triplex Triangle
as the Fibonacci and Jacobsthal sequences are to Pascal's Triangle
and the Duplex Triangle, respectively.*

The recursive rule is based on row 1 of the Triplex Triangle, in that it is the <u>string dot product</u> of the <u>reverse</u> of that row with two previous consecutive elements in the sequence. This gives us the following alternative version of the recursive rule:

$$S_n = (1, 3) \bullet (S_{n-1}, S_{n-2}) = 1 \cdot S_{n-1} + 3 \cdot S_{n-2}$$

In general, we have the following formulas, which are the Triplex extensions of the recursive $S_n$ Rule.

$$
\begin{aligned}
S_n = 1 \cdot S_n & \\
= 1 \cdot S_{n-1} + {} & 3 \cdot S_{n-2} & (26) \\
= 1 \cdot S_{n-2} + {} & 6 \cdot S_{n-3} + 9 \cdot S_{n-4} & (27) \\
= 1 \cdot S_{n-3} + {} & 9 \cdot S_{n-4} + 27 \cdot S_{n-5} + 27 \cdot S_{n-6} & (28) \\
= 1 \cdot S_{n-4} + {} & 12 \cdot S_{n-5} + 54 \cdot S_{n-6} + 108 \cdot S_{n-7} + 81 \cdot S_{n-8} & (29) \\
= 1 \cdot S_{n-5} + {} & 15 \cdot S_{n-6} + 90 \cdot S_{n-7} + 270 \cdot S_{n-8} + 405 \cdot S_{n-9} + 243 \cdot S_{n-10} & (30)
\end{aligned}
$$

etc.

We can also prove formula (27) by using formula (26), the given recursive rule for the $S_n$ sequence.

Given Formula (26): $S_n = 1 \cdot S_{n-1} + 3 \cdot S_{n-2}$
Use the recursive rule and replace $S_{n-1}$ and $S_{n-2}$ with
$1 \cdot S_{n-2} + 3 \cdot S_{n-3}$ and $1 \cdot S_{n-3} + 3 \cdot S_{n-4}$, respectively.

Therefore,
$$
\begin{aligned}
S_n & = 1 \cdot S_{n-1} + 3 \cdot S_{n-2} \\
& = 1 \cdot [1 \cdot S_{n-2} + 3 \cdot S_{n-3}] + 3 \cdot [1 \cdot S_{n-3} + 3 \cdot S_{n-4}] \\
& = 1 \cdot S_{n-2} + 3 \cdot S_{n-3} + 3 \cdot S_{n-3} + 9 \cdot S_{n-4} \\
& = 1 \cdot S_{n-2} + 6 \cdot S_{n-3} + 9 \cdot S_{n-4} \qquad \blacksquare
\end{aligned}
$$

### Section 6.12

1. Verify formula (27) for **(a)** $n = 5$ and **(b)** $n = 7$.
2. Verify formula (28) for **(a)** $n = 7$ and **(b)** $n = 8$.
3. Verify formula (29) for $n = 9$.
4. Prove formula (28) by using formula (26), the recursive rule for the $S_n$ sequence.
5. Verify formula (30) for $n = 10$.

6.     Write the string dot product using row 6 of the Triplex Triangle and the the appropriate sequence of consecutive $S_n$ numbers.

The $S_n$ recursive formula also fits nicely into another pattern of recursive formulas for $S_n$, see formulas (31) through (35).

$$S_n = 1 \cdot S_{n-1} + 3 \cdot 1 \cdot S_{n-2} = S_2 \cdot S_{n-1} + 3 \cdot S_1 \cdot S_{n-2}, \ n \geq 2 \tag{31}$$

$$= 3 \cdot S_{n-2} + 3 \cdot 1 \cdot S_{n-3} = S_3 \cdot S_{n-2} + 3 \cdot S_2 \cdot S_{n-3}, \ n \geq 3 \tag{32}$$

$$= 5 \cdot S_{n-3} + 3 \cdot 3 \cdot S_{n-4} = S_4 \cdot S_{n-3} + 3 \cdot S_3 \cdot S_{n-4}, \ n \geq 4 \tag{33}$$

$$= 11 \cdot S_{n-4} + 3 \cdot 5 \cdot S_{n-5} = S_5 \cdot S_{n-4} + 3 \cdot S_4 \cdot S_{n-5}, \ n \geq 5 \tag{34}$$

$$= 21 \cdot S_{n-5} + 3 \cdot 11 \cdot S_{n-6} = S_6 \cdot S_{n-5} + 3 \cdot S_5 \cdot S_{n-6}, \ n \geq 6 \tag{35}$$

etc.

Formulas (31) thru (35) are similar to formulas (15) thru (19) for the $R_n$ numbers in Section 6.9, except here the second term in the right member is multiplied by 3.

7.     Verify formula (33) for n = 6.
8.     Verify formula (34) for n = 7.
9.     Verify formula (35) for n = 6.
10.     Predict the next formula to follow formula (35).
11.     Prove formula (32) by using formula (31), the recursive rule for the $S_n$ sequence.

**In general, for given n and j, $j \leq n$,**

$$S_n = S_j \cdot S_{n-(j-1)} + 3 \cdot S_{j-1} \cdot S_{n-j}, \ n \geq j \tag{36}$$

12.     Let j = 7 and write out the formula (36).
13.     Verify formula (36) for n = 8 and j = 7.
14.     Let n = 2j − 1 in formula (36) and show that $S_{2j-1} = (S_j)^2 + 3 \cdot (S_{j-1})^2$.

Let j = n + 1 in formula (36), then a formula similar to formula (36) is

$$S_{2n} = S_{n+1} \cdot S_n + 3 \cdot S_n \cdot S_{n-1} = S_n \cdot (S_{n+1} + 3 \cdot S_{n-1}),$$

or equivalently,

$$\frac{S_{2n}}{S_n} = S_{n+1} + 3 \cdot S_{n-1}. \tag{37}$$

15.     Verify formula (37) for n = 2, 3, and 4.

16.    Verify the formula $S_{n+1} = [(S_n)^2 - (-3)^{n-1}] / S_{n-1}$ for $n = 2, 3,$ and 4.

17.    Verify the formula $S_n = \sum[C(n, j) \cdot 13^{(j-1)/2}] / (2^{n-1})$ (sum for $j = 1$ to n, but restrict j to just odd numbers) for $n = 2, 3, 4,$ and 5.

---

Consider the **rth** row of the Triplex Triangle (which consists of $r + 1$ elements), then choose $r + 1$ consecutive $S_n$ numbers from the $S_n$ sequence. Now form the **string dot product** of the string corresponding to the rth row of the Triplex Triangle and the string of $r + 1$ consecutive $S_n$ numbers. If the last $S_n$ number in the string equals $S_k$ (in position k) of the $S_n$ sequence, then the result of the product equals $S_{r+k}$.

**Example:** Choose row **3** and $3 + 1 = 4$ consecutive $S_n$ numbers starting with 4, that is, (4, 7, 19, 40). Then the last number is **40 = $S_6$** and is, therefore, in position **6**. Thus, $S_{3+6} = S_9 = 508$ is the result of the string dot product. The string for the 3$^{rd}$ row of the Triplex Triangle is (27, 27, 9, 1). The result of the string dot product is

$$(27, 27, 9, 1) \bullet (4, 7, 19, 40) = 27 \cdot 4 + 27 \cdot 7 + 9 \cdot 19 + 1 \cdot 40$$

$$= 108 + 189 + 171 + 40$$

$$= 508 = S_9.$$

---

16.    Form the string of the elements in row 2 of the Triplex Triangle and the string (4, 7, 19) of 3 consecutive elements of the $S_n$ sequence. Find the string dot product of these two strings and show that it is equal to $S_7 = 97$.

17.    Choose row 3 of the Triplex Triangle and $3 + 1 = 4$ consecutive $S_n$ numbers starting with $S_2 = 1$. Form the string dot product of the string for row 3 of the Triplex Triangle and the string of 4 consecutive $S_n$ numbers starting with $S_2 = 1$.
Verify that the position of last $S_n$ number in the string is in position $n = 5$ and that the result of the string dot product is equal to
$$S_{3+5} = S_8 = 217.$$

18.    Form the string consisting of $S_4$ and $S_5$. Use row 1 of the Triplex Triangle to form the second string and show that the string dot product of the two strings is equal to $S_6$.

---

## 6.13   The Sum of the $S_n$ Numbers

The sum of $S_n$ numbers from the first to the nth, $1 + 1 + 4 + 7 + 19 + 40 + \ldots + S_n$, is equal to one-third the quantity of $S_{n+2}$ minus 1. The formula is

$$1 + 1 + 4 + 7 + 19 + 40 + \ldots + S_n = (S_{n+2} - 1)/3 \qquad (38)$$

Chapter 5, Formula (22), gives the sum of the Jacobsthal numbers as

$$1 + 1 + 3 + 5 + 11 + 21 + \ldots + J_n = (J_{n+2} - 1)/2$$
$$= (J_{n+1} + 2 \cdot J_n - 1)/2.$$

By way of comparison, the sum of the $S_n$ numbers is given above as $(S_{n+2} - 1)/3$, but by the recursive rule, $S_{n+2}$ is equal to the sum of the previous $S_n$ number plus 3 times the $S_n$ number prior to that. Therefore, we can write

$$1 + 1 + 4 + 7 + 19 + 40 + \ldots + S_n = (S_{n+2} - 1)/3$$
$$= (S_{n+1} + 3 \cdot S_n - 1)/3. \quad (39)$$

**Section 6.13**

1.     Use the two formulas (38) and (39) to find the sum **1 + 1 + 4 + 7 + 19 + 40**. Verify that each formula yields the same result.

2.     Find the sum **1 + 1 + 4 + 7 + 19 + 40 + . . . + $S_n$** for n = 3, 5, 6 and 7 using one of the formulas above.

It is interesting to compare the sum of the Jacobsthal numbers and the sum of the $S_n$ numbers.

The sum of the Jacobsthal numbers $= (J_{n+2} - 1)/2 = (J_{n+1} + 2 \cdot J_n - 1)/2$.
The sum of the $S_n$ numbers $= (S_{n+2} - 1)/3 = (S_{n+1} + 3 \cdot S_n - 1)/3$.

## 6.14   A Sequence Related to the $S_n$ Numbers

We have a relationship with the Triplex Triangle and the $S_n$ numbers similar to the relationship we explored with the Duplex Triangle and the Jacobsthal diagonals in Chapter 5. We first look at the sum of $S_n$ numbers from $S_1$ to $S_n$ as given by formula (39), and that sum is equal to

$$(S_{n+2} - 1)/3.$$

Also, the sum of all the elements in the Triplex Triangle from row 0 to row n – 1 is equal to

$$(1/3)(4^n - 1)$$

(refer to Section 6.6).

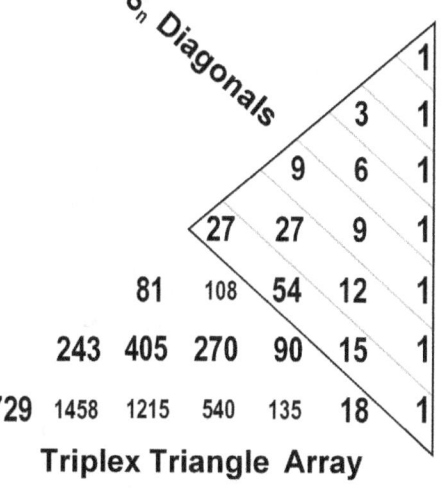

**Triplex Triangle Array**

Since the sum of the $S_n$ numbers correspond to elements enclosed by the triangle shown, the formula for the sum of the Triplex Triangle elements **not** represented by numbers on the $S_n$ diagonals for a given value of n, is equal to the sum of all the elements of the Triplex Triangle up to row $(n - 1)$ minus the sum of the $S_n$ numbers from $S_1$ to $S_n$.

That is, the sum of the Triplex Triangle elements through row $n - 1$ **not** represented by numbers on the $S_n$ diagonals for a given value of n, is equal to

$$(1/3)(4^n - 1) - (S_{n+2} - 1)/3 = (4^n - S_{n+2})/3 .$$

### Section 6.14

1.   For each of the figures below find **(a)** the sum of all of the elements shown in the Triplex Triangle, **(b)** the sum of the elements enclosed in the drawn triangle, and **(c)** the sum of the elements not enclosed in the triangle. In each case, compare your sums to the values of the expressions

   (1)   $(1/3)(4^n - 1)$ ,                    (2)   $(S_{n+2} - 1)/3$ ,

   (3)   $(1/3)(4^n - 1) - (S_{n+2} - 1)/3$,   and   (4)   $(4^n - S_{n+2})/3$ ,

   where n equals one more than the row number of the last row shown in each figure.

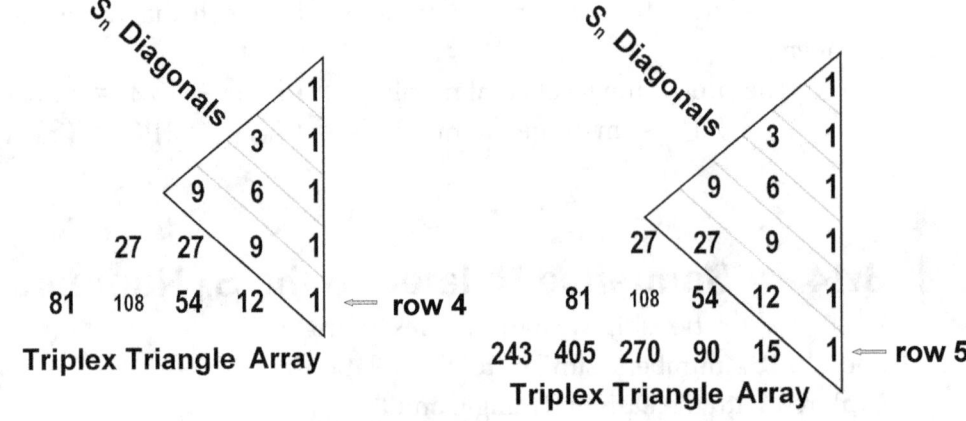

Triplex Triangle Array ← row 4

243  405  270  90  15  1  ← row 5
Triplex Triangle Array

2.   For the figure shown at the beginning of this section find **(a)** the sum of all of the elements shown in the Triplex Triangle, **(b)** the sum of the elements enclosed in the drawn triangle, and **(c)** the sum of the elements not enclosed in the triangle. In each case, compare your sums to the values of the expressions

   (1)   $(1/3)(4^n - 1)$ ,                    (2)   $(S_{n+2} - 1)/3$ ,

                       and   (3)   $(4^n - S_{n+2})/3$,

   where n equals one more than the row number of the last row shown in the figure.

The expression   $(4^n - S_{n+2})/3$   given above for the sum of those elements of the

Triplex Triangle that are **not enclosed** with the $S_n$ diagonals, provides a formula for a sequence of numbers which we will denote by

$$g(n) = (4^n - S_{n+2})/3, \text{ for } n = 1, 2, 3, \ldots.$$

This produces the sequence

**0, 3, 15, 72, 309, 1293, 5292, . . ., g(n), . . . .**

Also, notice that **g(4) = 72 = 4·g(3) + 12** and **g(5) = 309 = 4·g(4) + 21.** The 12 and 21 that were added in those two formulas are the numbers, **3·$S_3$** and **3·$S_4$.** Noting this we can write **g(4) = 4· g(3) + 3·$S_3$** and **g(5) = 4·g(4) + 3·$S_4$.**

---

3.    Complete the following table.  Compare the final values of columns 3 and 4 with the values in column 2.   Make a conjecture.

| n | g(n) | $(4^n - S_{n+2})/3$ | | $4 \cdot g(n-1) + 3 \cdot S_{n-1}$ | |
|---|------|---------------------|---|------------------------------------|---|
| 1 | 0 | $(4^1 - S_{1+2})/3$ = | 0 | $4 \cdot g(0) + 3 \cdot S_0$ = | 0 |
| 2 | 3 | $(4^2 - S_{2+2})/3$ = | 3 | $4 \cdot g(1) + 3 \cdot S_1$ = | 3 |
| 3 | 15 | $(4^3 - S_{3+2})/3$ = | 15 | $4 \cdot g(2) + 3 \cdot S_2$ = | 15 |
| 4 | 72 | $(4^4 - S_{4+2})/3$ = | 72 | $4 \cdot g(3) + 3 \cdot S_3$ = | 72 |
| 5 | 309 | $(4^5 - S_{5+2})/3$ = _____ | | $4 \cdot g(4) + 3 \cdot S_4$ = _____ | |
| 6 | 1293 | $(4^6 - S_{6+2})/3$ = _____ | | $4 \cdot g(5) + 3 \cdot S_5$ = _____ | |
| 7 | 5292 | _____ = _____ | | _____ = _____ | |
| 8 | ____ | _____ = _____ | | _____ = _____ | |

---

In general,

$$g(n) = (4^n - S_{n+2})/3 = 4 \cdot g(n-1) + 3 \cdot S_{n-1}. \tag{40}$$

Another formula for the general term of the sequence **g(n)** is

$$g(n) = g(n-1) + 4^{n-1} - S_n. \tag{41}$$

---

4.    Evaluate formula (41) for n = 3, 4, and 5.  Verify the values found with those in the table above.

It is interesting to note the resemblance of the formula for **a(n)**, the sequence associated with Pascal's Triangle and the Fibonacci diagonals and the formula for **b(n)**, the sequence associated with the Duplex Triangle and the Pell diagonals (Chapter 4), and the formula for **c(n)**, the sequence associated with the Duplex Triangle and the Jacobsthal anti-diagonals (Chapter 5), and the formula for **e(n)**, the sequence associated with the Triplex Triangle and the $R_n$ diagonals. Compare those with the formula for **g(n)**.

$$a(n) = 2 \cdot a(n-1) + F_{n-1},$$

anti-diagonals

$$b(n) = 3 \cdot b(n-1) + P_{n-1}, \quad c(n) = 3 \cdot c(n-1) + 2 \cdot J_{n-1},$$

$$e(n) = 4 \cdot e(n-1) + R_{n-1}, \quad g(n) = 4 \cdot g(n-1) + 3 \cdot S_{n-1}.$$

## 6.15   The $S_n$ – Pythagorean Triples

Recall the sides of a right triangle, **(a, b, c)**, are determined by **Euclid's formulas:**

$$a = x^2 - y^2, \quad b = 2xy, \quad \text{and} \quad c = x^2 + y^2$$

where **x > y** and **a** and **b** are the legs of the right triangle and **c** is the hypotenuse, so that **a, b,** and **c** satisfy the Pythagorean formula:

$$c^2 = a^2 + b^2.$$

Now if we choose consecutive $S_n$ numbers for x and y, then **c**, the hypotenuse, will be a function of $S_n$ numbers. Examine the table below:

| x | y | $a = x^2 - y^2$ | $b = 2xy$ | $c = x^2 + y^2$ |
|---|---|---|---|---|
| $S_2$ = 1 | 1 | 0 | 2 | $2 = S_3 - 2 \cdot (S_1)^2$ ⟸ row 1 |
| $S_3$ = 4 | 1 | 15 | 8 | $17 = S_5 - 2 \cdot (S_2)^2$ |
| $S_4$ = 7 | 4 | 33 | 56 | $65 = S_7 - 2 \cdot (S_3)^2$ |
| $S_5$ = 19 | 7 | 312 | 266 | $410 = S_9 - 2 \cdot (S_4)^2$ |
| $S_6$ = 40 | 19 | 1239 | 1520 | $1961 = S_{11} - 2 \cdot (S_5)^2$ |

. . .   . . .   *(row 1 does not correspond to a triangle)*

Notice that the subscripts of the first $S_n$ numbers in the last column are odd numbers and are equal to the sum of the subscripts of $S_n$ numbers represented by x and y. Note that the difference, **c – b,** is the square of three times a $S_n$ number. Also note that the sum, **c + b,** is the square of the numbers in the sequence

$$2, 5, 11, 26, 59, \ldots, s_n, \ldots,$$

where $s_n = s_{n-1} + 3 \cdot s_{n-2}$ for n > 2 and $s_1 = 2, s_2 = 5$. We will call this sequence the "**companion** $S_n$ **sequence**".

### Section 6.15

1. Extend the table above by forming the next line in the sequence.
2. Verify that $S_{13} - 2 \cdot (S_5)^2$ is equal to the value of **c** in your extension of the table.
3. Verify that the difference, **c – b**, in each line of the table involving the $S_n$ numbers, is the square of three times a $S_n$ number.
4. Verify that the sum, **c + b**, in each line of the table involving the $S_n$ numbers, is the square of a number in the sequence called the "companion $S_n$ sequence", that is, 2, 5, 11, 26, 59, ..., $s_n$, ....
5. Show that the first few elements of the sequence called the "companion $S_n$ sequence", that is, 2, 5, 11, 26, 59, ..., $s_n$, ..., are equal to sums of two consecutive $S_n$ numbers.
6. For each row of the table involving the $S_n$ numbers, show that
   **(a)** the perimeter, **a + b + c,** for that row is equal to $2 \cdot S_{n+1} \cdot s_n$, and
   **(b)** the area is equal to $(3 \cdot S_{n-1} \cdot S_n)/2$ times the perimeter.

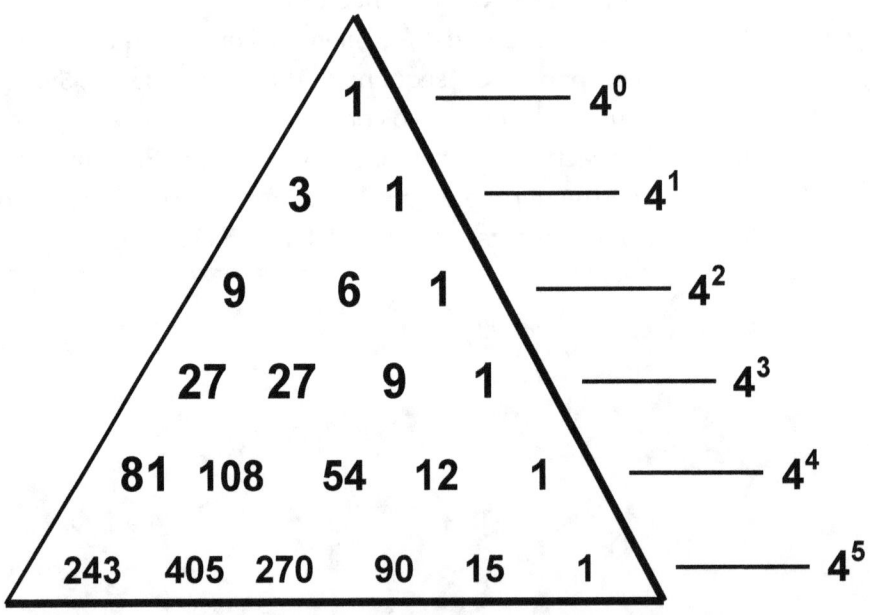

**Triplex Triangle**

# 7

# A Family of Pascal Related Triangles

## 7.1 The Pascal Triangle Family of Triangles

In the previous chapters we saw how the expansion of the binomial $(n + 1)^d$ leads to triangular arrays. The elements of Pascal's Triangle (and the Simplex Triangle) correspond to the coefficients in the expansion when n = 1. The elements of the Duplex Triangle correspond to the coefficients in the expansion when n = 2 and the elements of of the Triplex Triangle correspond to the coefficients in the expansion when n = 3. For these triangular arrays we have the following equations that develop elements in the rows of the triangles for each value of **d = 0, 1, 2, 3, . . . .**

$$(1 + 1)^d = \sum C(d, k) \quad\quad = 2^d, \text{ for } k = 0 \text{ to } d, \quad \textbf{(Pascal, Simplex)}$$

$$(2 + 1)^d = \sum C(d, k) \cdot 2^{d-k} = 3^d, \text{ for } k = 0 \text{ to } d, \quad \textbf{(Duplex)}$$

$$(3 + 1)^d = \sum C(d, k) \cdot 3^{d-k} = 4^d, \text{ for } k = 0 \text{ to } d \quad \textbf{(Triplex).}$$

The next equation in the sequence,

$$(4 + 1)^d = \sum C(d, k) \cdot 4^{d-k} = 5^d, \text{ for } k = 0 \text{ to } d,$$

also leads to a triangular array that will have many properties similar to the properties we explored in the previous chapters. Therefore, in this chapter, and any chapters that might follow, you, the reader, are asked to write the details of the chapter.

Look for the row sums, the recursive rule for the elements in the subsequent rows and extensions of the recursive rules, alternating row sums and various number patterns within the columns and cross-columns. You can find a formula for the sum of all of the rows of the triangle up to row n, that is, $5^0 + 5^1 + 5^2 + . . . + 5^d$. There are number patterns related to string dot products of strings formed from rows and columns of this triangle and other triangles, possibly including Vandermonde's Identity. Then there are the diagonals and anti-diagonals to consider in finding new sequences with their own recursive rules, properties and related sequences. And perhaps new findings related to the Pythagorean connection.

1. Develop the triangular array corresponding to the equation
   $$(4 + 1)^d = \sum C(d, k) \cdot 4^{d-k} = 5^d, \quad \text{for } k = 0 \text{ to } d,$$
   and find out as much as you can about it.  Use the triangular array below to get you started.

2. Continue to develop the triangular arrays corresponding to the equations that follow from expanding $(n + 1)^d$ for subsequent values of n.

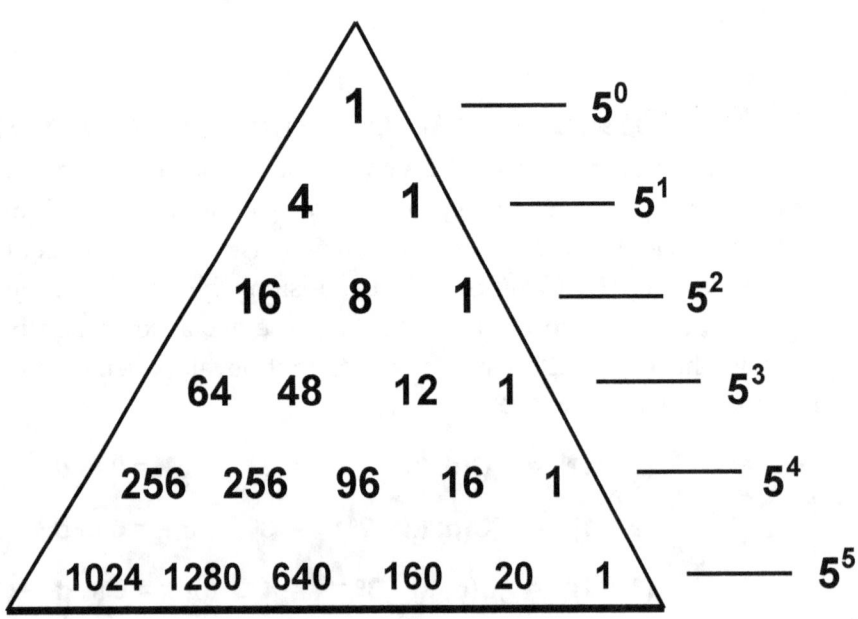

**Quadplex Triangle**

**NOT THE END**

- There may be no end to the discoveries you can make.

# NUMBER SEQUENCES

**Triangular Numbers** (2nd column Pascal's Triangle, partial sums of natural numbers)
1, 3, 6, 10, 15, 21, 28, 36, 45, 55, 66, 78, 91, 105, 120, 136, 153, 171, 190, . . .

**Square Numbers** ( $n^2$ )
1, 4, 9, 16, 25, 36, 49, 64, 81, 100, 121, 144, 169, 196, 225, 256, 289, 324, . . .

**Cubic Numbers** ( $n^3$ )
1, 8, 27, 64, 125, 216, 343, 512, 729, 1000, 1331, 1728, 2197, 2744, 3375, . . .

**Powers of 2**
1, 2, 4, 8, 16, 32, 64, 128, 256, 512, 1024, 2048, 4096, 8192, 16384, . . .

**Powers of 3**
1, 3, 9, 27, 81, 243, 729, 2187, 6561, 19683, 59049, 177147, . . .

**Powers of 4**
1, 4, 16, 64, 256, 1024, 4096, 16384, 65536, 262144, 1 048 576, . . .

**Tetrahedral Numbers** (or triangular pyramidal, 3rd column Pascal's Triangle)
1, 4, 10, 20, 35, 56, 84, 120, 165, 220, 286, 364, 455, 560, 680, 816, 969, 1140, . . .

**Pentatopal Numbers** (4th column Pascal's Triangle, 4-simplexal)
1, 5, 15, 35, 70, 126, 210, 330, 495, 715, 1001, 1365, 1820, 2380, 3060, 3876, . . .

**5-D Simplex Numbers** (5th column Pascal's Triangle, 5-simplexal)
1, 6, 21, 56, 126, 252, 462, 792, 1287, 2002, 3003, 4368, 6188, 8568, 11626, . . .

**Fibonacci Numbers** ( $F_n = F_{n-1} + F_{n-2}$, where $F_1 = 1$ and $F_2 = 1$)
1, 1, 2, 3, 5, 8, 13, 21, 34, 55, 89, 144, 233, 377, 610, 987, 1597, 2584, 4181, 6765, . . .

**Lucas Numbers** ( $L_n = L_{n-1} + L_{n-2}$, where $L_1 = 1$ and $L_2 = 3$)
1, 3, 4, 7, 11, 18, 29, 47, 76, 123, 199, 322, 521, 843, 1364, 2207, 3571, 5778, 9349, . . .

**Pell Number Sequence** (Diagonal sums of the Duplex triangle)
0, 1, 2, 5, 12, 29, 70, 169, 408, 985, . . . , $P_n$ , . . .
where $n = 0, 1, 2, 3, \ldots$ and $P_n = 2 \cdot P_{n-1} + P_{n-2}$ for $n = 2, 3, \ldots$ and $P_0 = 0$ and $P_1 = 1$.

**Companion Pell Numbers**
2, 2, 6, 14, 34, 82, 198, 478, 1154, . . . , $Q_n$ , . . .
where $n = 0, 1, 2, 3, \ldots$ and $Q_n = 2 \cdot Q_{n-1} + Q_{n-2}$ for $n = 2, 3, \ldots$ and $Q_0 = 2$ and $Q_1 = 2$.

**Jacobsthal Number Sequence** (Anti-Diagonal sums of the Duplex triangle)
0, 1, 1, 3, 5, 11, 21, 43, 85, 171, 341, . . . , $J_n$ , . . .
where $n = 0, 1, 2, 3, \ldots$ and $J_n = 1 \cdot J_{n-1} + 2 \cdot J_{n-2}$ for $n = 2, 3, \ldots$ and $J_0 = 0$ and $J_1 = 1$.

**Companion Jacobsthal numbers**
2, 1, 5, 7, 17, 31, 65, 127, 257, 511, 1025, . . . , $K_n$ , . . .
where $n = 0, 1, 2, 3, \ldots$ and $K_n = K_{n-1} + 2 \cdot K_{n-2}$ for $n = 2, 3, \ldots$ and $K_0 = 2$ and $K_1 = 1$.

**Sequence of Diagonal sums of the Triplex triangle**
    1, 3, 10, 33, 109, 360, 1189, . . . , $R_n$, . . .
    where n = 1, 2, 3, . . . and $R_n = 3 \cdot R_{n-1} + R_{n-2}$ for n = 2, 3, . . . and $R_0 = 0$ and $R_1 = 1$.

**Sequence of Anti-Diagonal sums of the Triplex triangle**
    1, 1, 4, 7, 19, 40, 97, . . ., $S_n$ , . . .
    where n = 1, 2, 3, . . . and $S_n = S_{n-1} + 3 \cdot S_{n-2}$ for n = 2, 3, . . . and $S_0 = 0$ and $S_1 = 1$.

**The Sequence a(n) associated with Pascal's Triangle and the Fibonacci sequence**
    0, 1, 3, 8, 19, 43, 94, 201, . . ., a(n), . . . ,
    where n = 1, 2, 3, . . . and $a(n) = 2^n - F_{n+2} = 2 \cdot a(n-1) + F_{n-1}$.

**The Sequence b(n) associated with the Duplex Triangle and the Pell sequence**
    0, 1, 5, 20, 72, 245, 805, 2584, . . ., b(n), . . . .
    where n = 1, 2, 3, . . . and $b(n) = 3 \cdot b(n-1) + P_{n-1}$.

**The Sequence c(n) associated with the Duplex Triangle and the Jacobsthal sequence**
    0, 2, 8, 30, 100, 322, 1008, 3110, . . ., c(n), . . .
    where n = 1, 2, 3, . . . and $c(n) = (3^n - J_{n+2})/2 = 3 \cdot c(n-1) + 2 \cdot J_{n-1}$.

**The Sequence e(n) associated with the Triplex Triangle and the $R_n$ sequence**
    0, 1, 7, 38, 185, 849, 3756, . . ., e(n), . . .
    where n = 1, 2, 3, . . . and $e(n) = 4 \cdot e(n-1) + R_{n-1}$ for n = 2, 3, . . . and e(0) = 0 and e(1) = 0.

**The Sequence g(n) associated with the Triplex Triangle and the $S_n$ sequence**
    0, 3, 15, 72, 309, 1293, 5292, . . ., g(n), . . . .
    where n = 1, 2, 3, . . . and $g(n) = 4 \cdot g(n-1) + 3 \cdot S_{n-1}$ for n = 2, 3, . . . and g(0) = 0 and g(1) = 0.

**The Sequence of Partial Sums of the Fibonacci Numbers**
    1, 2, 4, 7, 12, 20, 33, 54, 88, 143, . . . , $G_n$ , . . .
    where n = 1, 2, 3, . . . and $G_n = G_{n-1} + G_{n-2} + 1$ for n = 3, 4, . . . and $G_1 = 1$ and $G_2 = 2$.

**The Sequence of Partial Sums of the Jacobsthal Numbers**
    1, 2, 5, 10, 21, 42, 85, 170, 341, 682, . . . , $H_n$ , . . .
    where n = 1, 2, 3, . . . and $H_n = H_{n-1} + 2 \cdot H_{n-2} + 1$ for n = 3, 4, . . . and $H_1 = 1$ and $H_2 = 2$.

**The Sequence of Partial Sums of the Pell Numbers**
    1, 3, 8, 20, 49, 119, 288, 696, . . . , $T_n$ , . . .
    where n = 1, 2, 3, . . . and $T_n = 2 \cdot T_{n-1} + 1 \cdot T_{n-2} + 1$ for n = 3, 4, . . . and $T_1 = 1$ and $T_2 = 3$.

**The Sequence of Partial Sums of the $R_n$ Numbers**
    1, 4, 14, 47, 156, 516, 1705, . . . , $U_n$ , . . .
    where n = 1, 2, 3, . . . and $U_n = 3 \cdot U_{n-1} + 1 \cdot U_{n-2} + 1$ and $U_1 = 1$ and $U_2 = 4$.

**The Sequence of Partial Sums of the $S_n$ Numbers**

1, 2, 6, 13, 32, 72, 169, 386, 894, . . . , $V_n$ , . . .

where $n = 1, 2, 3,$ . . . and $V_n = V_{n-1} + 3 \cdot V_{n-2} + 1$ and $V_1 = 1$ and $V_2 = 2$.

**Sequence of Partial Sums of the Powers of 2**

1, 3, 7, 15, 31, 63, 127, 255, . . . , $II_n$ , . . .

where $n = 1, 2, 3,$ . . . and $II_n = 2 \cdot II_{n-1} + 1$ for n= 2, 3, . . . and $II_1 = 1$,

or

1, 3, 7, 15, 31, 63, 127, 255, . . . , $2^{n+1} - 1$, . . .

for $n = 0, 1, 2, 3,$ . . . .

**Sequence of Partial Sums of the Powers of 3**

1, 4, 13, 40, 121, 364, 1093. . . , $III_n$ , . . .

where $n = 1, 2, 3,$ . . . and $III_n = 3 \cdot III_{n-1} + 1$ for n= 2, 3, . . . and $III_1 = 1$,

or

1, 4, 13, 40, 121, 364, 1093. . . , $(1/2)(3^{n+1} - 1)$, . . .

for $n = 0, 1, 2, 3,$ . . . .

**Sequence of Partial Sums of the Powers of 4**

1, 5, 21, 85, 341, 1365, . . . , $J_{2n+2}$, . . .

for $n = 0, 1, 2, 3,$ . . . and $J_{2n+2}$ is a Jacobsthal number,

or

1, 5, 21, 85, 341, 1365, . . . , $IV_n$ , . . .

where $n = 1, 2, 3,$ . . . and $IV_n = 4 \cdot IV_{n-1} + 1$ for n= 2, 3, . . . and $IV_1 = 1$,

or

1, 5, 21, 85, 341, 1365, . . . , $(1/3)(4^{n+1} - 1)$ , . . .

for $n = 0, 1, 2, 3,$ . . . .

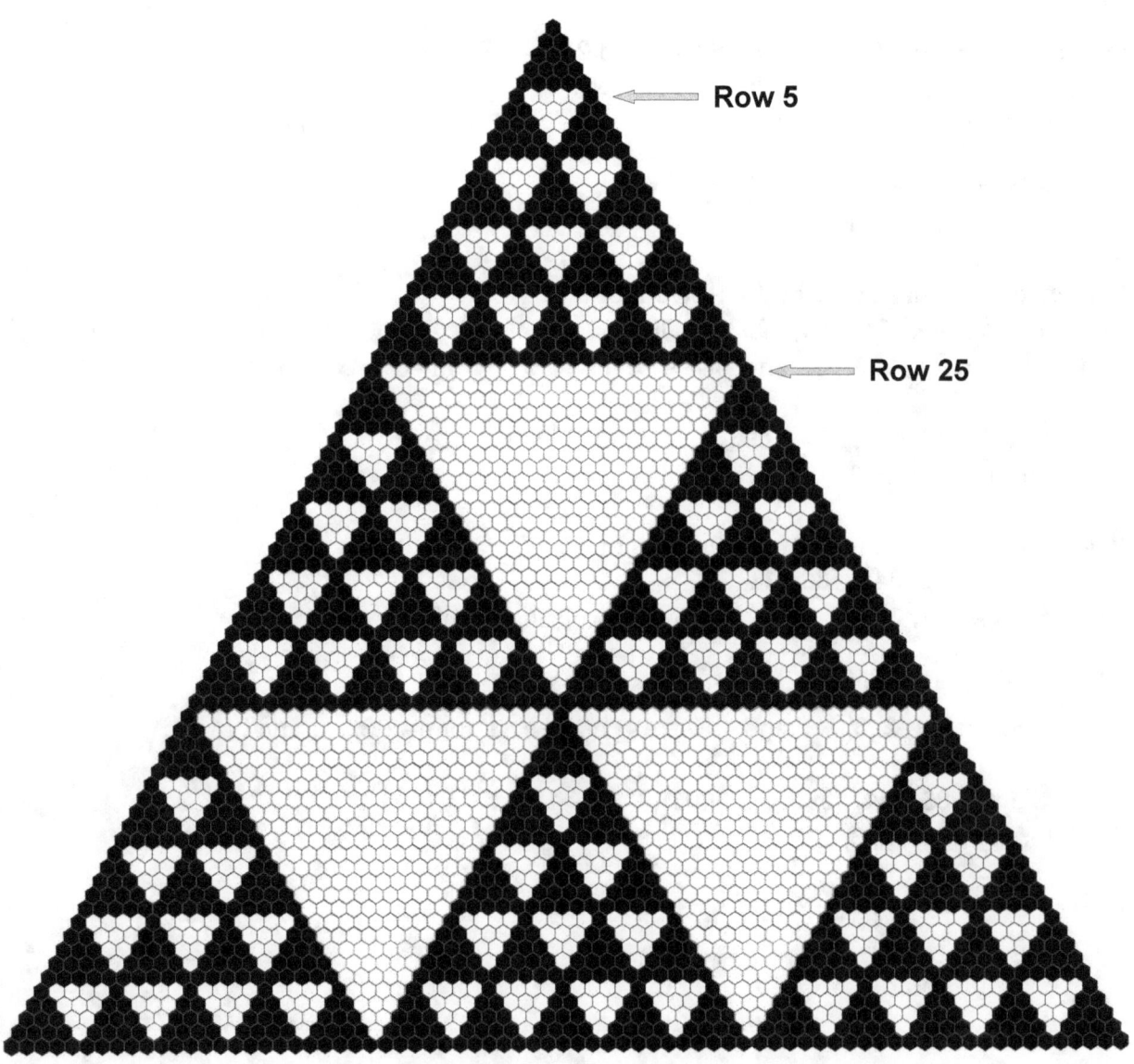

Row 5

Row 25

**Multiples of 5 are the white elements
in this Pascal Triangle.**

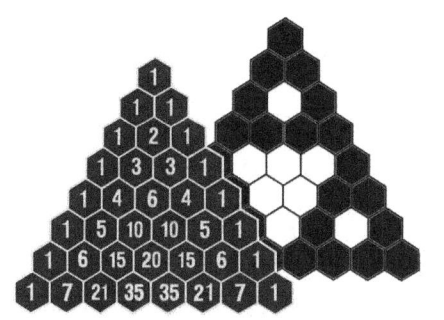

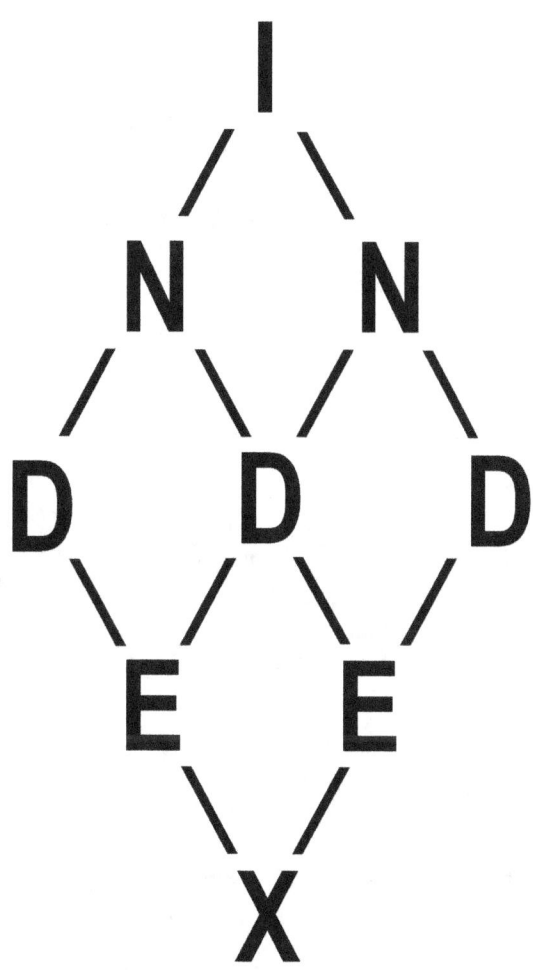

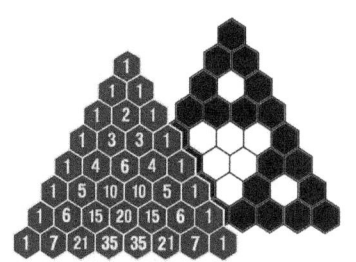

## About the Author

Thomas M. Green was born in New York State, raised in Pennsylvania, Texas, and California. He received his Master's Degree in Mathematics in 1962 from California State University, Long Beach. He did post-graduate study at Stanford University in the late 1960's and early 1970's as a fellow in two different National Science Foundation projects in Mathematics and Computer Science.

He has authored or co-authored several mathematical papers, textbooks, the book *Pascal's Triangle, 1986* (with Charles L. Hamberg), *Pascal's Triangle, 2nd Ed., 2013* and the book *Texas Hold 'Em Poker Textbook, 2010*.

Member of the Mathematical Association of America, Tom is now professor emeritus of Mathematics at Contra Costa College. He lives in California with his wife Sandra.

www.ingramcontent.com/pod-product-compliance
Lightning Source LLC
Chambersburg PA
CBHW080809180526
45168CB00006B/2382